C0-AKG-396

INTERNATIONAL SERIES OF MONOGRAPHS IN

ANALYTICAL CHEMISTRY

General Editors: R. Belcher and H. Freiser

VOLUME 55

NATURAL CHELATING POLYMERS

NATURAL CHELATING POLYMERS

ALGINIC ACID, CHITIN AND CHITOSAN

R. A. A. MUZZARELLI

Professor of Quantitative Analytical Chemistry and
Professor of Applied Radiochemistry,
University of Bologna,
Bologna, Italy

PERGAMON PRESS
OXFORD · NEW YORK · TORONTO
SYDNEY · BRAUNSCHWEIG

Pergamon Press Ltd., Headington Hill Hall, Oxford
Pergamon Press Inc., Maxwell House, Fairview Park, Elmsford, New York 10523
Pergamon of Canada Ltd., 207 Queen's Quay West, Toronto 1
Pergamon Press (Aust.) Pty. Ltd., 19a Boundary Street, Rushcutters Bay, N.S.W. 2011, Australia
Vieweg & Sohn GmbH, Burgplatz 1, Braunschweig

First edition 1973

Library of Congress Cataloging in Publication Data

Muzzarelli, R A A 1937-
Natural chelating polymers.

(International series of monographs in analytical chemistry, v. 55)
1. Chemical tests and reagents. 2. Chelates. 3. Algin. 4. Chitin. 5. Chitosan. I. Title.
QD77.M84 1973 547'.84 73-3402
ISBN 0-08-017235-0

Printed in Hungary

CONTENTS

PREFACE

THE development of new materials should be accompanied by both evolution of thinking modes and improvement of working methods: too often a research field is considered exhausted, while a more thorough study of the information reached and an intelligent look at parallel research fields would possibly reveal connections of data.

Today, in certain fields of research, academic, military and industrial activities are progressing with little exchange of information: for instance, the International Symposium on Ion Exchange in London in 1969 was unattended by University and Agency researchers, while the International Symposium on Analytical Chemistry at Birmingham the same year was unattended by industrial men.

In a few countries much is being done for the widest and best utilization of the scientific information: however, after reading this book everybody will be convinced that the existing information on alginic, chitin and chitosan was not used in a modern way, and it remained confined to specialized teams working in many different disciplines, with no exchanges.

May I cite a few facts whose documentation will be found in the appropriate place in this book:

(i) alginic acid has the same backbone and functional groups as carboxymethylcellulose, but it was never mentioned in papers dealing with the latter, and vice versa;
(ii) chitin has the same backbone and a functional group similar to aminated celluloses, but it was never mentioned in papers dealing with the latter, and vice versa;
(iii) authors failed to see these simple connections, just because the polymers mentioned belong to traditionally "different" research

projects, to the point that on carboxymethylcellulose coupling with dithizone, a U.S. Patent was issued and a paper on a leading journal in analytical chemistry was published without any mention or reference to alginic acid;

(iv) in spite of the large industrial applications of alginic acid and its derivatives and salts, it is not clear today which action these polymers exert on the trace transition elements present in the human body;

(v) chitin and chitosan were analyzed with many different techniques: elemental analysis, infrared and ultraviolet spectrometry, X-ray diffractions, nuclear magnetic resonance and so on, but nobody checked the phosphorous, silicon and transition element content. Nobody cared about "metals" as they were doing "biology", "entomology" or "physiology", while "metals" traditionally belong to other fields. Today we know that minute amounts of certain metals can significantly alter the spectra;

(vi) alginic acid was extensively used in the form of biscuits for radiocontamination protection studies on rats and swine, while chitosan was fed to rats and mice for investigating its anti-ulcer properties; nevertheless, chitosan has not yet been tested for radiocontamination protection;

(vii) agarose could be prepared by using diethylaminoethylcellulose, according to a U.S. Patent issued in 1969: however, the inventor did not mention chitin and chitosan which in 1971 was shown by other authors to be superior to diethylaminoethylcellulose for making agarose;

(viii) dozens of fine papers have been published on the rather obvious topic that resins carrying ethylenediaminetetraacetic acid groups exhibit a complexing capacity similar to ethylenediaminetetraacetic acid solutions, while until 1968 nobody cared to observe or report how deep blue chitosan becomes on reacting with copper ions:[1] it is interesting to recall that chitosan was isolated 74 years earlier.

Fundamentals of ion-exchange cellulose design and usage in biochemistry were recently reviewed, but no mention to ligand exchange chromatography or to metal ion-treated celluloses was made.[2] My early applications of modified celluloses in inorganic chromatography[3] led to the demonstration of the complexing ability of this class of polymers, particularly diethylaminoethyl, aminoethyl, para-aminobenzyl, ecteola and guanidoethyl celluloses. I have written a chapter on this subject to gather together the existing information on cellulose column chromatography of inorganic compounds,[4] and it was published in 1967: reviewers unanimously predicted popularity for this topic.[5–8]

In fact the 1972 catalogue of an important supplier lists twenty-seven different celluloses for research and industrial applications. The 1971 Guide to Literature on Whatman Advanced Ion-Exchange Celluloses[9] includes 322 papers, all of them dated 1968–70, and this emphasizes the wide applicability of these new celluloses as chromatographic supports. Nevertheless, in the above-mentioned Guide papers dealing with metal ions are omitted, in view of its specialized character; therefore as a number of papers have been recently published on inorganic chromatography on celluloses, the most significant of them will be mentioned briefly below.

What is most important is that these studies on celluloses led me to search for new, improved performance substances, by an original approach: isolation from natural products instead of synthesis or modification. About the same time I proposed chitin and chitosan as natural chelating polymers containing amino groups, another Italian team proposed alginic acid as a natural chelating polymer containing carboxyl groups: our research activity came to our reciprocal attention at a meeting sponsored by the National Research Council of Italy, in Rome in 1969.

A striking fact is, above all, that chelating polymers can be obtained from the enormous amount of natural raw materials that are presently discharged to sea as waste from the marine canning factories. This is now changing, however, as a result of serious pollution problems caused by the annual dumping of thousand of tons of lobster, shrimp and crab carcasses, a material which is highly resistant to biodegradation. Recognition of this situation prompted the National Oceanic Atmospheric Administration, Office of Sea Grant, U.S. Department of Commerce, to support a research programme at the University of Washington and broaden the efforts to seek and develop economically sound ways of utilizing the waste products from marine food processing plants. Therefore, the task of collecting some bibliographical material was undertaken, and a bibliography of 593 selected publications on chitin was published with the intention of "stimulating the initiation of new theoretical and applied research projects in areas that await exploration".[10] Moreover, a pilot plant is now producing chitin and chitosan of consistent, controllable quality.

Many applications are foreseen for chelating polymers, and I am confident that this book will be of use to analytical chemists as well as to scientists engaged in other fields.

I acknowledge with thanks the financial assistance of the National Research Council of Italy and of the International Atomic Energy Agency of Vienna, which made possible the production of the original data presented in this book, and the preparation of the manuscript.

REFERENCES

1. R. A. A. Muzzarelli and O. Tubertini, *Talanta* **16,** 1571 (1969); see also U.S. Patent 3, 635, 818 (1972).
2. C. S. Knight, Some fundamentals of ion-exchange cellulose design and usage, in *Advances in Chromatography*, Vol. IV (R. A. Keller and J. C. Giddings, Eds.), Marcel Dekker, New York, 1967.
3. R. A. A. Muzzarelli, Applications of radioisotopes in chromatography on substituted celluloses, I—Cobalt. *Talanta* **13,** 193 (1966).
4. R. A. A. Muzzarelli, Inorganic chromatography on columns of natural and substituted celluloses, in *Advances in Chromatography*, Vol. V (R. A. Keller and J. C. Giddings, Eds.), Marcel Dekker, New York, 1967.
5. *Nature* **219,** 657 (1968).
6. *Chem. Britain* **4,** 510 (1968).
7. *Anal. Chem.* **40,** 53 R (1968).
8. *Analyst* **93,** 835 (1968).
9. *Advanced Ion-Exchange Celluloses, Guide to Literature*, W. & R. Balston Ltd. Maidstone, 1971.
10. E. R. Pariser and S. Bock, *Chitin and Chitin Derivatives, An Annotated Bibliography from 1965 through 1971.* Rept. No MITSG 73–2, M.I.T., Cambridge (15 Oct. 1972).

CHAPTER 1

RECENT APPLICATIONS OF MODIFIED CELLULOSES IN INORGANIC ANALYTICAL CHEMISTRY

Cellulose can be used in certain cases to collect metal ions, particularly when organic solvents are used, for instance, traces of antimony could be separated from gram amounts of iron and manganese in ethyl–ether-nitric acid mixture, as reported in Fig. 1.1, as well as from gram amounts of mercuric chloride, uranyl nitrate, and tetrachloroauric acid.[(1)] Under similar conditions, nanograms of mercury could be separated from milligrams of bismuth, as presented in Fig. 1.2;[(2)] uranium in large amounts of uranyl nitrate dissolved in ethyl ether could be separated from many metals, among which are those shown in Fig. 1.3.[(3)]

Modified celluloses can collect metal ions from solutions much more effectively than natural cellulose.[(4)] Therefore, attempts to attach carboxyl groups or amino groups to the backbone of cellulose have been carried out in recent years by many authors.[(5–14)]

However, it is a well established fact that by chemical modification no more than 1·5% nitrogen can be introduced in the polymer, by chemical means. In one case higher values have been reported.[(15)] In Fig. 1.4 the nitrogen per cent of diethylaminohydroxypropylcellulose DEAHPCell is indicated as a function of diethylepoxypropylamine DEEPA concentration in the reacting mixture, and values as high as 3·8% are reached after 1 hr. This product can be obtained after treating cellulose with 16% alkali and keeping it wet: should it dry, less nitrogen would be fixed, because the nitrogen percentage is a function of the DEEPA/H_2O ratios.

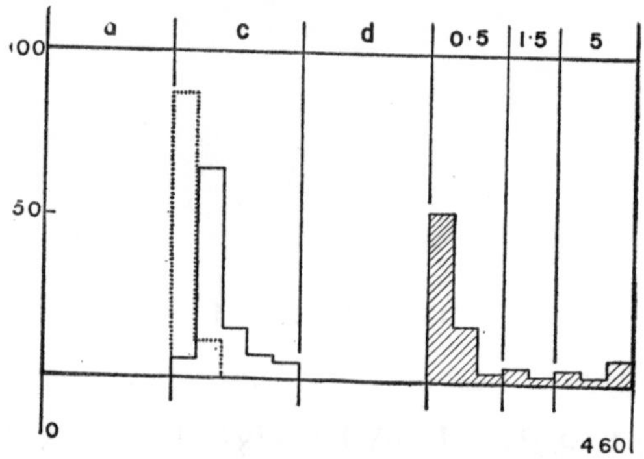

FIG. 1.1. The chromatographic separation of 30 nanograms of antimony (shaded area), from gram amounts of iron (.....) and manganese (——) on a column of natural cellulose (15×1 cm). Flow-rate 20 ml min^{-1}. (From R. A. A. Muzzarelli *et al.*, *Talanta* **14,** 305 (1967).)

a = anhydrous ethyl ether. c = 40 mg ammonium thiocyanate in 20 ml methanol + 80 ml ethyl ether. d = 5 g of ammonium thiocyanate in 20 ml methanol + 80 ml ethyl ether, and hydrochloric acid in methanol.

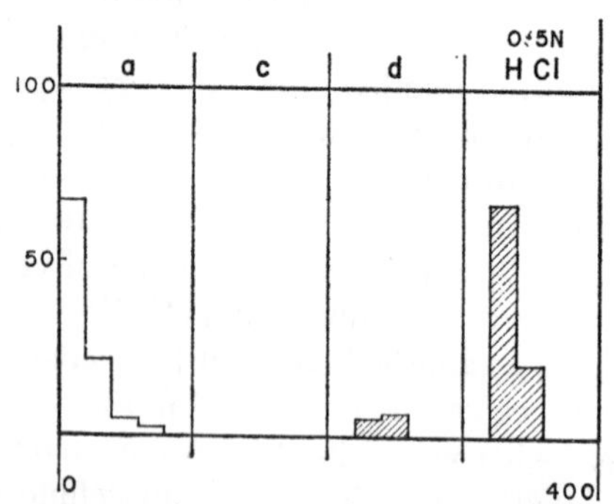

FIG. 1.2. The separation of nanograms of mercury (——) from milligrams of bismuth (shaded area) on a column of natural cellulose (15×1 cm). Flow-rate 20 ml min^{-1}. (From R. A. A. Muzzarelli *et al.*, *Talanta* **14,** 489 (1967).)

a = anhydrous ethyl ether. c = 40 mg ammonium thiocyanate in 20 ml methanol + 80 ml ethyl ether. d = 5 g of ammonium thiocyanate in 20 ml methanol + 80 ml ethyl ether, and hydrochloric acid in methanol.

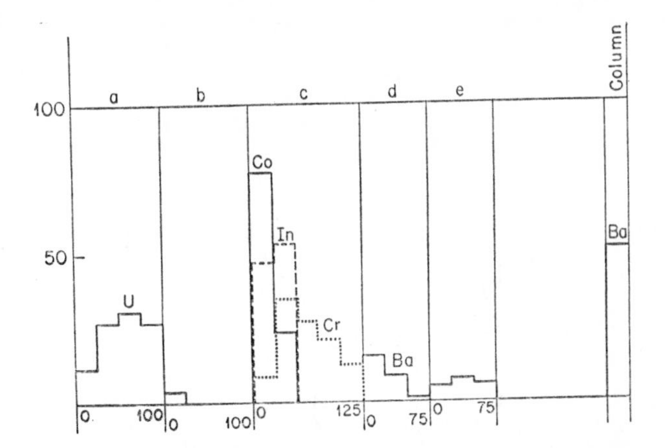

FIG. 1.3. The chromatographic separation of gram amounts of uranyl nitrate from trace amounts of other elements, on a 20×1 cm column of cellulose in ethyl ether. (From R. A. A. Muzzarelli *et al.*, *Talanta* **12,** 823 (1965).)

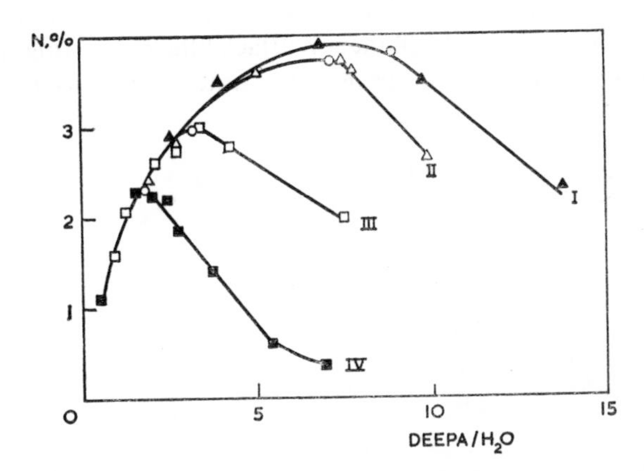

FIG. 1.4. Influence of the weight composition of the amine mixture on the nitrogen percent of the products. $T = 100$ °C, I is for DEEPA/Cell = 6·2, II = 5·0, III = 2·2, and IV = 1·2. (From R. M. Noreika, *Cellul. Chem. Tech.* **5,** 117 (1971).)

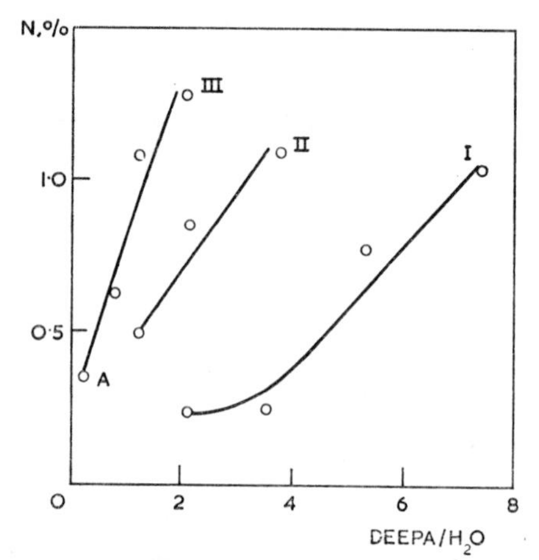

FIG. 1.5. Influence of various treatment procedures on the nitrogen percent of the products. I = cellulose in DEEPA to which the calculated amount of water was added later, with DEEPA/H_2O = 13, II = cellulose in DEEPA+H_2O 7 : 1 mixture previously prepared; III = cellulose in excess of DEEPA+H_2O 7 : 1 mixture. T=100 °C, time 120 min. (From R. M. Noreika, *Cellul. Chem. Tech.* **5,** 117 (1971).)

The preparation procedure itself has a marked effect on the yield, as in Fig. 1.5.

The mechanism of the reaction includes the protonation of the amino group, the cellulosate nucleophilic ion attacks the epoxy ring, to be later neutralized by a proton-rich species, like water or cellulose.

$$\underset{\text{O}}{\text{CH}_2\!-\!\text{CH}}\!-\!\text{CH}_2\!-\!\text{N}(\text{C}_2\text{H}_5)_2 + \text{H}_2\text{O} \rightleftharpoons \underset{\text{O}}{\text{CH}_2\!-\!\text{CH}}\!-\!\text{CH}_2\!-\!\text{N}(\text{C}_2\text{H}_5)_2 \ldots \text{HOH} \qquad (1)$$

$$\underset{\text{O}}{\text{CH}_2\!-\!\text{CH}}\!-\!\text{CH}_2\!-\!\text{N}(\text{C}_2\text{H}_5)_2 \ldots \text{HOH} \rightleftharpoons \underset{\text{O}}{\text{CH}_2\!-\!\text{CH}}\!-\!\text{CH}_2\!-\!{}^{\oplus}\text{N}(\text{C}_2\text{H}_5)_2\!:\!\text{H} + {}^{\ominus}\text{OH} \qquad (2)$$

$$\text{CellOH} + \text{OH}^{\ominus} \rightleftharpoons \text{CellO}^{\ominus} + \text{H}_2\text{O} \qquad (3)$$

$$\text{Cell O}^{\ominus} + \underset{\text{O}}{CH_2\text{—}CH}\text{—}CH_2\text{—}N(C_2H_5)_2 \rightarrow \text{Cell O—}CH_2\text{—}\underset{O^{\ominus}}{CH}\text{—}CH_2\text{—}N(C_2H_5)_2 \xrightarrow{+HOR} \text{Cell O—}CH_2\text{—}\underset{OH}{CH}\text{—}CH_2\text{—}N(C_2H_5)_2 + {}^{\ominus}OR \qquad (4)$$

For diethylaminoethylcellulose DECell, other authors have proposed mechanisms where the ionization of cellulose is not included. Of course the preparation of the derivatives of cellulose is beyond the scope of this book and the above information is reported as an example.

If we compare the chromatographic behaviour of copper on natural cellulose and on amino substituted cellulose, as in Fig. 1.6, we obtain a clear indication of a different mode of interaction mostly due to the presence of amino groups on the polymer.[16]

From Figs. 1.6 to 1.8, as well as from other data reported in the original papers, it is evident that the capacity of substituted celluloses is higher

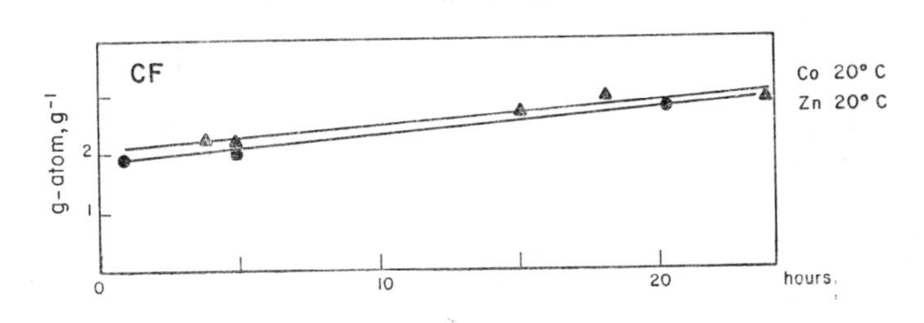

FIG. 1.6. The collection of zinc and cobalt nitrates hexahydrates in ethyl ether, on natural cellulose, (10^{-6} g atoms of metal per gram of cellulose vs. hours). 50 ml of a 0·25 mM solution with 1 g of cellulose. ▲ = cobalt, ● = zinc. $T = 20$ °C. (From R. A. A. Muzzarelli *et al.*, *Anal. Chem.* **39,** 1762 (1967).)

than that of natural cellulose. The collection is also quite rapid taking place during the first hour contact time, and more metal is collected during the next 24 hr by the substituted celluloses.

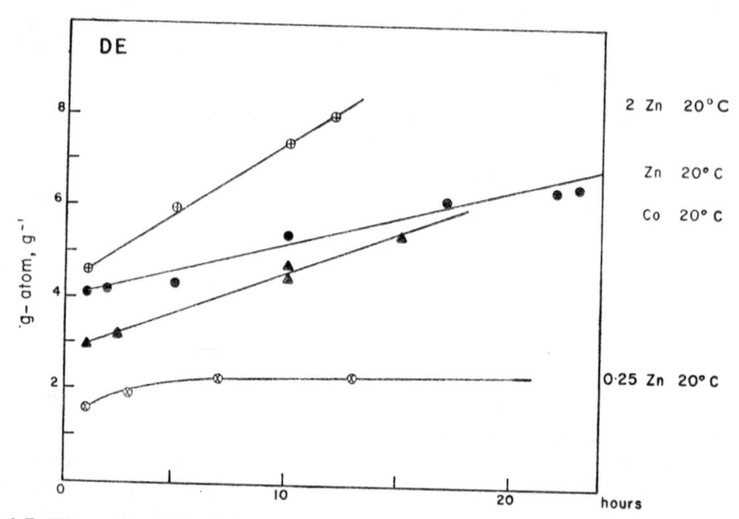

FIG. 1.7. The collection of zinc and cobalt nitrates hexahydrates in ethyl ether, on diethylaminoethylcellulose (DE). (10^{-6} g atoms of metal per gram of DE cellulose vs. hours. 50 ml of 0·50, 0·25 or 0·06 mM solutions with 1 g DE cellulose. (From R. A. A. Muzzarelli *et al., Anal. Chem.* **39,** 1762 (1967).)

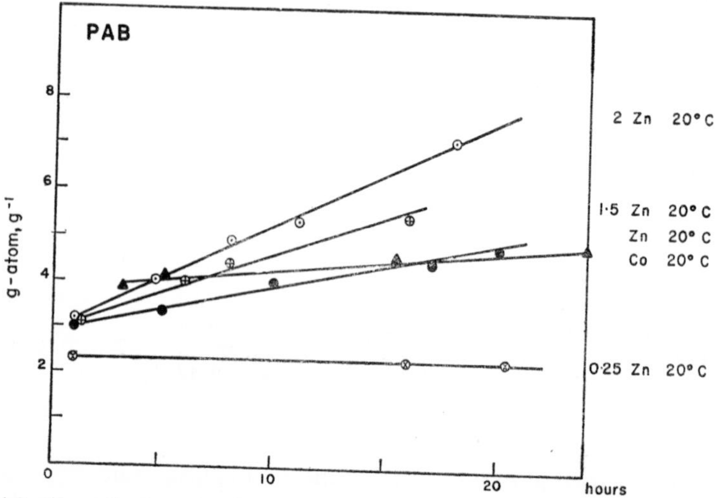

FIG. 1.8. The collection of zinc and cobalt nitrates hexahydrates in ethyl ether, on paraaminobenzylcellulose (PAB). (Conditions and reference as in Fig. 1.7.)

The comparative analysis of the data under standard conditions revealed that all celluloses, including natural cellulose, rapidly collect 2 μg-atom of zinc or copper $gram^{-1}$, and some of them collect much more due to their complexing ability.

In aqueous solutions, substituted celluloses can collect transition elements to a rather limited extent, but in a few instances they are very effective. Data obtained under reference conditions for the most important polymers mentioned in this book are presented in seven tables: the first two tables are Tables 1.1 and 1.2 where data are reported for DE and PAB celluloses in aqueous solutions. To facilitate the comparison each table contains data relevant to two polymers.

In Table 1.1 it can be seen that, under the same conditions, DE cellulose is more effective than PAB cellulose in collecting first-row transition metal ions. However, it can be noticed that values for 12 hr contact time in most cases are lower than the corresponding values at 1 hr. DE cellulose can collect much better the metals listed in Table 1.2, especially molybdate

TABLE 1.1. FIRST-ROW TRANSITION AND POST-TRANSITION METAL ION COLLECTION ON 200 mg DIETHYLAMINOETHYLCELLULOSE (DE) AND PARAAMINOBENZYLCELLULOSE (PAB). PER CENT OF THE AMOUNT OF METAL PRESENT IN 50 ML OF 0·44 mM AQUEOUS SOLUTION (ATOMIC ABSORPTION SPECTROMETRY)

	pH		hr	Cr(III)	Cr(VI)	Mn(II)	Fe(II)	Ni(II)	Cu(II)	Zn(II)	As(V)
DE	2·5		1	58	85	8	14	4	3	0	10
			12	13	80	2	0	30	31	10	100
		EDTA	1	20	41	7	6	10	0	17	0
	5·5		1	65	89	33	71	27	81	86	90
			12		98	13	38	26	86	50	100
		EDTA	1	21	19	22	0	14	0	0	4
PAB	2·5		1	5	38	3	3	0	0	0	0
			12								
		EDTA	1	21	12	0	6	0	0	0	0
	5·5		1	30	33	15	19	0	12	16	0
			12								
		EDTA	1	5	11	19	2	10	0	0	0

(R. A. A. Muzzarelli, original results.)

TABLE 1.2. SECOND- AND THIRD-ROW TRANSITION AND POST-TRANSITION METAL ION COLLECTION ON 200 mg DIETHYLAMINOETHYLCELLULOSE (DE) AND PARA-AMINOBENZYLCELLULOSE (PAB). PER CENT OF THE AMOUNT PRESENT IN 50 ml OF 0·44 mM AQUEOUS SOLUTION (ATOMIC ABSORPTION SPECTROMETRIY)

	pH		hr	Mo(VI)	Ag(I)	Sn(II)	Sb tartrate	Hg(II)	Pb(II)
DE	2·5		1	100	23		37	72	5
			12	100	97	100		72	0
		EDTA	1		15		6	32	0
	5·5		1	100	100			30	86
			12	100	100			72	98
		EDTA	1					35	0
PAB	2·5		1	83	27		11	7	0
			12						
		EDTA	1		16		2	8	
	5·5		1	87				21	24
			12						
		EDTA	1					0	0

(R. A. A. Muzzarelli, original results.)

silver and mercury. It can be said that PAB is very selective for molybdate, and this finding was very important for establishing a new method for the analytical determination of Mo in sea-water, as described on pages 199–201.

The above researches include a final elution of certain elements with methanol–hydrochloric acid mixture. Recent works on the collection of a number of metals on DE cellulose, in methanol–hydrochloric acid mixture, extended the information to other metals.[17] It was found that zinc, cadmium, mercury and bismuth are strongly held on DE as shown in Table 1.3: the data for bismuth confirms its chromatographic behaviour and its high K_d value is in agreement with the difficulty encountered in eluting it with methanol–hydrochloric acid. The presence of methanol allows zirconium, tellurium and thallium to be slightly collected on DE. Most other metals do not adsorb on DE from this medium, and exhibit K_d values lower than ten: they are: Be, Mg, Ca, Sr, Ba, Sc, Y, Lantha-

TABLE 1.3. DISTRIBUTION COEFFICIENTS ON DE CELLULOSE

Metal	[HCl], M	Methanol, % v/v				
		95·2	90·4	80·9	71·4	50·0
Zn	1	$8\cdot3\times10^{2}$	$3\cdot7\times10^{2}$	$1\cdot8\times10^{2}$	76	31
	3	$8\cdot3\times10^{2}$	$3\cdot6\times10^{2}$	$1\cdot1\times10^{2}$	47	27
	6	$3\cdot9\times10^{2}$	$2\cdot2\times10^{2}$	73	35	13
	12	$3\cdot9\times10^{2}$	$1\cdot8\times10^{2}$	62	30	11
Cd	1	*	$1\cdot9\times10^{3}$	$5\cdot6\times10^{2}$	$3\cdot6\times10^{2}$	68
	3	*	$9\cdot7\times10^{2}$	$2\cdot6\times10^{2}$	$1\cdot6\times10^{2}$	32
	6	$1\cdot2\times10^{3}$	$4\cdot0\times10^{2}$	$1\cdot2\times10^{2}$	$1\cdot6\times10^{2}$	30
	9	$1\cdot0\times10^{3}$	$3\cdot6\times10^{2}$	84	34	14
	12	$8\cdot7\times10^{2}$	$2\cdot2\times10^{2}$	58	24	11
Hg(II)	1	*	*	$7\cdot3\times10^{2}$	$5\cdot8\times10^{2}$	
	3	*	$1\cdot9\times10^{3}$	$4\cdot3\times10^{2}$	$1\cdot8\times10^{2}$	
	6	*	$1\cdot2\times10^{3}$	$1\cdot5\times10^{2}$	$1\cdot1\times10^{2}$	
	9	*	$8\cdot7\times10^{2}$	$1\cdot3\times10^{2}$	$1\cdot0\times10^{2}$	
	12	$1\cdot6\times10^{3}$	$3\cdot3\times10^{2}$	75	14	
Bi	1	*	*	*	$1\cdot4\times10^{3}$	$7\cdot2\times10^{2}$
	3	*	*	*	$9\cdot6\times10^{2}$	$1\cdot9\times10^{2}$
	6	*	*	*	$4\cdot8\times10^{2}$	$1\cdot0\times10^{2}$
	9	*	*	$1\cdot1\times10^{3}$	$3\cdot3\times10^{2}$	65
	12	*	*	$9\cdot1\times10^{2}$	$3\cdot3\times10^{2}$	32

* K_d greater than 2×10^{3}.
Loading: Zn 0·19, Cd 0·093, Hg(II) 0·012 and Bi 0·093 meq for 1 g of DE.

(From R. Kuroda *et al.*, *Talanta* **18,** 1123 (1971).)

nons, Ti(III and IV), Hf V(VI), Cr(III), Mo(VI), Mn(II), Re(VII), Fe(III), Co(II), Ni, Cu, Al, Ga, In, Sn(IV), As(III), Sb(III), Se(IV) and U(VI). Among the platinum group elements, Pd, Ir(IV) and Pt(IV) are strongly collected on DE from methanol–hydrochloric acid, while Ru(III) and Rh(II) are collected to a lower extent. Although the addition of methanol greatly assists the formation of chlorocomplexes and causes the reduction of concentration of competitive chloride ions, thus promoting the adsorption of chlorocomplexes on DE, few metals are sufficiently strongly adsorbed to permit their separation on the columns. Each of the four metals reported in Table 1.3, were separated from about 100 mg each of eighteen metals including Mg, Ca, La, Sm, Th, U(VI), Mo(VI) Mn(II), Fe(III), Co(II), Ni, Cu, In, Sn(IV), As(III), Sb(III); Se(IV) and Te(IV).

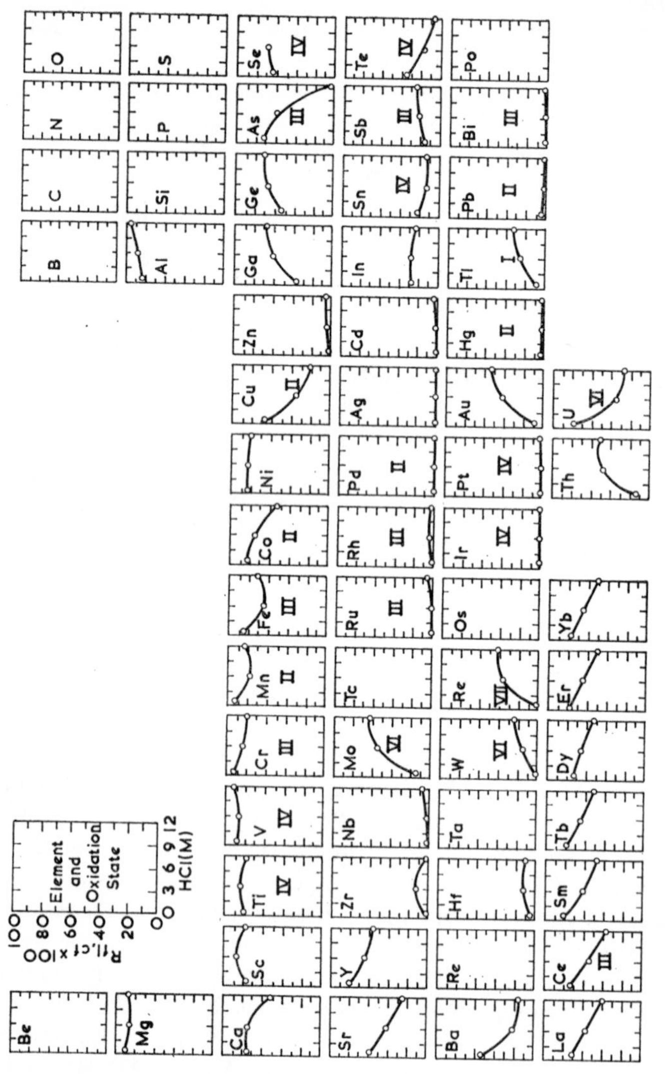

FIG. 1.9. R_f spectra of metals chromatographed on DE cellulose in methanol–HCl (20 : 1). (From R. Kuroda *et al.*, *J. Chromatogr.* **47**, 453 (1970).)

It might be surprising to notice that in methanol–hydrochloric acid mixture mercury has a high K_d value and it is strongly collected on DE cellulose,[17] while in ethyl ether mercuric chloride, iodide and nitrate are not collected to any extent.[18] It might be possible that in the presence of hydrochloric acid some sort of insoluble compound involving the amino groups of DE cellulose is formed, like the familiar insoluble amino-mercury(II) chloride, while in ether and in the absence of hydrochloric

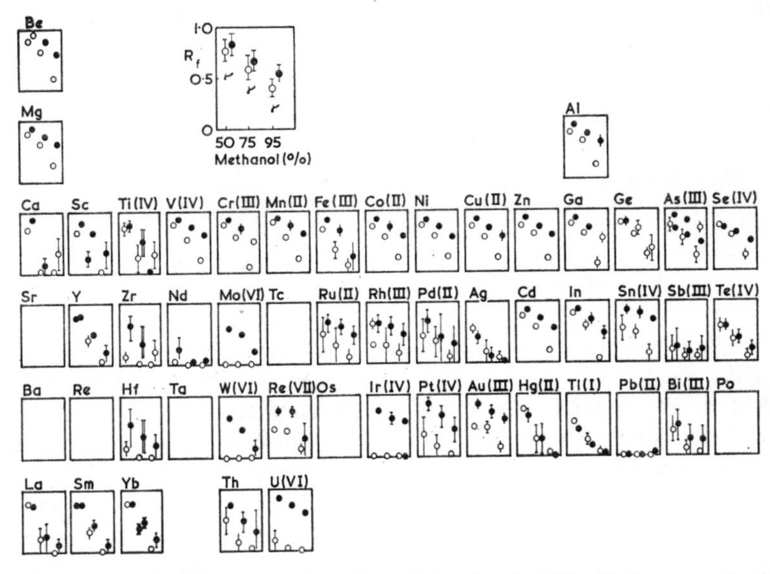

FIG. 1.10. Variation of R_f values of metals on DE and Avicel SF with the concentration of methanol in methanol–1 N H_2SO_4 media. ○ = DE; ● = Avicel SF. All the R_f measurements were conducted at 50, 75 and 95% methanol concentrations. For convenience R_f values on Avicel SF shifted the abscissa arbitrarily to the right. (From K. Oguma *et al.*, *J. Chromatogr.* **61,** 307 (1971).)

acid the complexation of mercury by the amino groups of cellulose is prevented.

Similar results, in Figs. 1.9 and 1.10 were reported for thin layers of DE cellulose in methanol with hydrochloric[19] or sulphuric acid[20] respectively.

Zinc, cadmium and bismuth show no marked tendency toward collection on DE in aqueous thiocyanate-chloride media, while mercury (II)

TABLE 1.4. EFFECT OF TOTAL CHLORIDE CONCENTRATION ON THE DISTRIBUTION COEFFICIENT OF MERCURY(II)

pH	Conc. of Cl^-						
	0	0·010	0·030	0·10	0·30	0·50	1·0
With DE –SCN^- form*							
3·0†	$>10^4$	$>10^4$	$>10^4$	$2·8\times10^3$	$3·0\times10^2$	$1·2\times10^2$	34
1·0‡				$2·9\times10^3$	$3·3\times10^2$	$1·2\times10^2$	28
With DE –Cl^+ form**							
3·0†	$9·1\times10^2$	$1·1\times10^3$	$1·2\times10^3$	$6·8\times10^2$	$2·0\times10^2$	87	27
1·0‡				$6·3\times10^2$	$2·2\times10^2$	95	28

* The concentration of NH_4SCN was kept at 0·010 M throughout.

† KCl was added to pH 3 acetate buffer solution to give the chloride concentration listed.

‡ To KCl solution of the listed chloride concentration. HCl solution of the same chloride concentration was added to give the chloride buffer solution of pH 1.

** No thiocyanate present.

(From R. Kuroda *et al.*, *Anal. Chim. Acta* **40**, 305 (1968).)

does: it was reported in fact, that mercury can be separated from many elements by passing a solution on a DE cellulose column in thiocyanate form.(21) This seems to be a precipitation of insoluble mercury thiocyanate, as it is well known that, among thiocyanates, mercury thiocyanate only is insoluble. In fact, data in Table 1.4 shows that mercury has no tendency to be collected on DE when in other chemical forms. The authors do not discuss the separation mechanism, and it might be possible that the same separation is feasible on any other support containing SCN^- ions, as it seems that the amino groups of DE cellulose in this case provide only basicity to hold in place SCN^- ions which in turn precipitate mercury(II) thiocyanate.

As in the methanol–ether mixtures used by the author, ammonium thiocyanate is a general complexing agent in aqueous solutions as well, but for antimony it was observed in all cases that thiocyanate is not effective.

The special behaviour of platinum and palladium was also observed, and it was reported that molybdenum and tungsten require 0·1 M NaOH +0·1 M NaCl for stripping.(21) This information is of interest as a background to the findings for chitin and chitosan.

In aqueous HCNS, R_f values have been determined for a number of metal ions on aminoethyl and DE cellulose papers, as well as on unmodified cellulose. The trends are the same as on column and thin layers, i.e. DE is the most effective in interacting with ions, all metals are prevented from fixing on the polymer in a thiocyanate medium, excepted mercury,

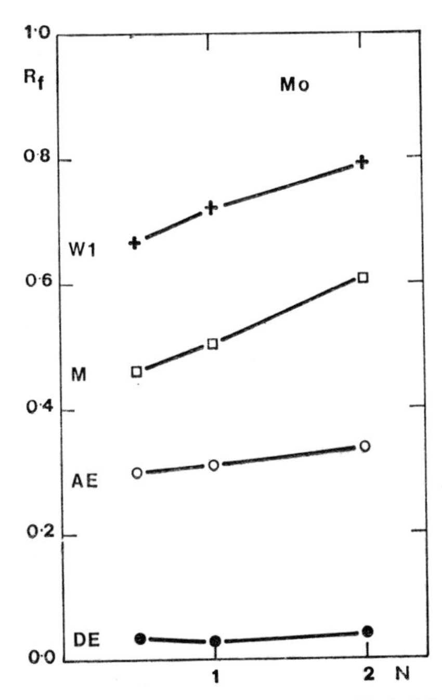

FIG. 1.11. R_f values of molybdenum on Whatman 1 (W1), Macherey & Nagel (M) aminoethylcellulose (AE) and diethylaminoethylcellulose (DE) papers with 0·5, 1·0 and 2·0 N HCNS as eluent. (From G. Bagliano *et al.*, *J. Chromatogr.* **21**, 471 (1966).)

indium, bismuth, iron(III) and zinc, and the behaviour of metavanadate, molybdate and tungstate is like in the other instances, as shown in Fig. 1.11,[22] in agreement with data of the other authors.

DE cellulose is particularly effective in collecting molybdenum and tungsten from thiocyanate solutions, as shown in Fig. 1.12.[23] Collection is at a maximum when pH is between 3·0 and 4·0. Therefore a mixture of perrhenate, tungstate and molybdate was submitted to chromatography

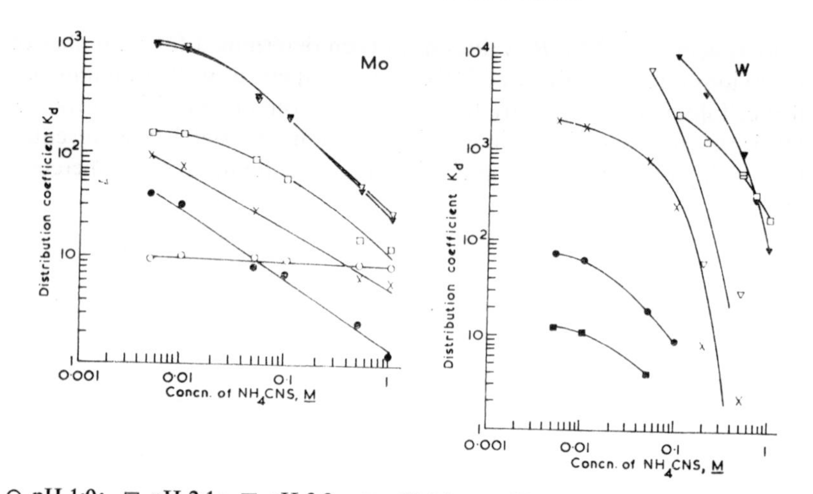

○ pH 1·0: □ pH 2·1: ▼ pH 3·2: ▽ pH 4·0: × pH 5·0: ● pH 6·0: ■ pH 6·9.

FIG. 1.12. Adsorption of Mo(VI) and W(VI) on DE cellulose from thiocyanate solutions. (From K. Ishida *et al.*, *Anal. Chem.* **39**, 212 (1967).)

on a column of DE cellulose in SCN form, and the metals were determined spectrophotometrically in the effluents. The sample solution was 0·02 M in NH_4CNS at pH 3·0, and it was added to the column at the flow-rate of 0·6 ml/min^{-1}. As can be observed in Fig. 1.13, rhenium is not collected, only appreciably retarded, while molybdenum and tungsten are collected; they can be selectively eluted from the DE–SCN column with 0·1 M NH-SCN at pH = 5·0 and 0·1 M NaOH + NaCl respectively. Various other anions had no effect on this separation, and results were presented for a mixture of the three metal ions in different proportions. Those authors did not report any information or comment about the mechanism of this separation, but in the opinion of the author the collection of molybdenum and tungsten on DE cellulose at those pH values is due to a particular interaction of these oxyanions with the functional groups of the polymer at pH values where homo- or eteropolyderivatives can be formed. This interaction is rather weak with respect to a similar interaction with chitosan, as expected by the lower number of amino-groups available on DE, but apparently it is of the same type. It is interesting to note that rhenium, which does not form poly derivatives, is not collected, as is the case for other transition elements. Chromium-

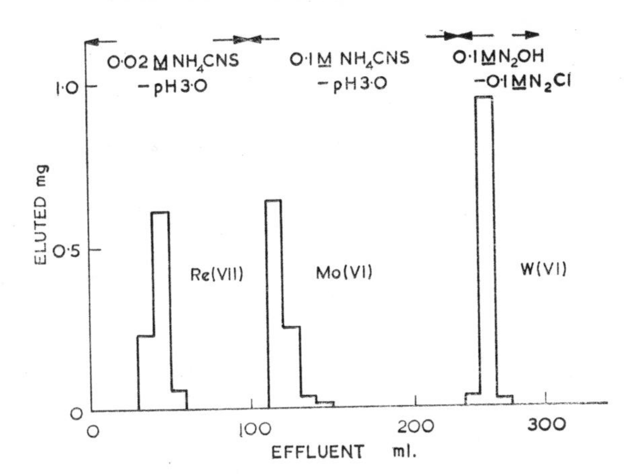

FIG. 1.13. Chromatographic separation of Re(VII), Mo(VI) and W(VI). (From K. Ishida *et al., Anal. Chem.* **39**, 212 (1967).)

(III) thiocyanate complexes on DE cellulose have been studied and reciprocally separated,[24] but the operating conditions are too different in the two works. Data on the behaviour of molybdenum in paper partition chromatography in thiocyanic acid indicate high R_f values at pH 3·0–3·9.[25] Therefore, as the paper was natural cellulose, this indicates the essential role of the diethylaminoethyl groups of DE cellulose in retaining molybdenum from thiocyanate solutions at low pH. Further, the capacity of modified celluloses was reported in Fig. 1.14 to decrease with pH, i.e. the amino group assumes the quaternary form which prevents interaction with metal ions; therefore, as molybdenum and tungsten are collected even from acidic thiocyanate solutions, their interaction with DE cellulose is a quite important and exceptional reaction.

DE cellulose was studied from the standpoint of its interactions with organic molecules too. Albumin can be collected on DE in the chloride form according to the Langmuir isotherm, and on DE in the hydroxide form according to the linear isotherm of sorption. The data obtained was interpreted on the basis of the theory of ion-exchange sorption of ampholytic organic compounds and it was calculated that the number of bonds which fix the protein molecule to the polymer is between 10 and 100.[26]

Among other studies, it is of interest to cite a method for making agarose based on the use of DE Cellulose, the ion-exchanger being character-

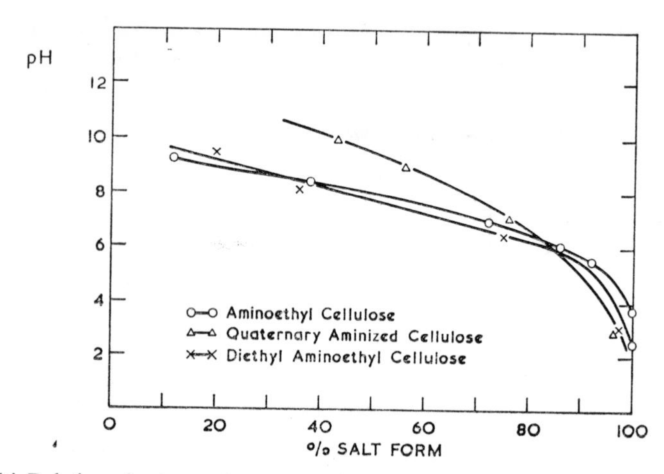

FIG. 1.14. Relation of anion-exchange capacity to pH in chemically modified celluloses. (From C. L. Hoffpauir *et al.*, *Text. Res. J.* **20**, 617 (1950).)

TABLE 1.5. COLLECTION OF AMINES ON METAL ION PRE-TREATED CELLULOSES (μmoles g^{-1}) AFTER SHAKING FOR 1 HR 0·1 g OF CELLULOSE IN 50 ml OF ETHER WITH 12 μmoles PER 50 ml

	CF				DE			
	Untreated	Co	Ag	Sb	Untreated	Co	Ag	Sb
Trimethylamine	49	49	76	49	25	70	88	40
Dimethylamine	29	55	70	60	25	65	70	65
Ethylenediamine	completely adsorbed				completely adsorbed			
Aniline	6	2	0	17	14	9	0	8
	P				PAB			
	Untreated	Co	Ag	Sb	Untreated	Co	Ag	Sb
Trimethylamine	40	58	66	69	65	75	78	59
Dimethylamine	106	55	101	101	71	55	44	100
Ethylenediamine	completely adsorbed				completely adsorbed			
Aniline	9	0	0	0	0	9	13	0

Each value is the average of at least three measurements.
Precision: ±8 per cent on ten measurements.

(From R. A. A. Muzzarelli *et al.*, *Analyst* **94**, 616 (1969).)

ized by pores large enough to admit molecules having a molecular weight of 25,000 or more, and the eluate obtained from the ion-exchanger containing agarose substantially free from agaropectin. The agaropectin content of agar is retained by DE cellulose.[27]

As far as ligand-exchange chromatography is concerned, it was found out that substituted celluloses carrying metal ions are effective in collecting amines.[28] From the data presented in Table 1.5, it is quite clear that the presence of the metal ion adds much to the capacity of celluloses for amine collection. In correspondence with the sequence listed in Table 1.5 for amine collection on DE cellulose treated with antimony, the separation of the four amines was carried out as shown in Fig. 1.15. The fixation

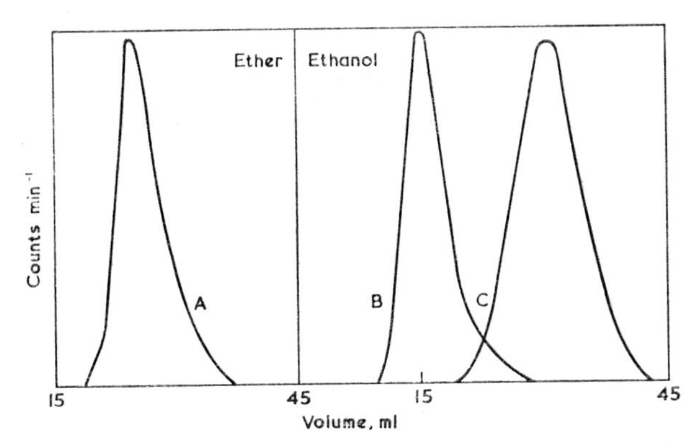

FIG. 1.15. Chromatographic behaviour of A, aniline; B, trimethylamine; and C, dimethylamine on antimony pre-treated diethylaminoethylcellulose DE; column 15×1 cm in ether. These amines are also separated from ethylenediamine, which is retained on the column. (From R. A. A. Muzzarelli *et al.*, *Analyst* **94**, 616 (1969).)

of the metal ion, the amine or both together appreciably altered the X-ray diffraction spectrum of the celluloses, according to the data in Table 1.6.

Shimomura obtained also results with phosphate cellulose, however, in water solutions, metal ion treated celluloses may present a metal ion leakage.[29]

Among the very many attempts to modify celluloses in view of better performances in collecting metal ions from aqueous solutions, the studies of Tolmachev and of Bauman can be cited: their approach is to attach to the modified cellulose some complexing molecule, and to produce a

TABLE 1.6. X-RAY DIFFRACTION DATA OF METAL ION PRE-TREATED CELLULOSES: 2θ

	h, k, l	Untreated	+Dimethyl amine	+Cobalt	+Cobalt +dimethyl amine*	+Anti-mony	+Anti mony +dimethyl-amine*
CF	(0,0,2)	22° 40′	22° 40′	22° 22′	22° 33′	22° 42′	22° 40′
	(1,0,1)	16° 24′	16° 20′	16° 12′	16° 30′	16° 26′	16° 23′
	(1,0,1)	14° 50′	14° 42′	14° 34′	14° 40′	14° 47′	14° 49′
DE	(0,0,2)	22° 37′	22° 30′	22° 21′	22° 35′	22° 38′	22° 36′
	( )	20° 03′	20° 00′	19° 52′	20° 00′	20° 00′	20° 00′
	(1,0,1)	16° 27′	16° 24′	16° 12′	16° 12′	16° 30′	16° 26′
	(1,0,1)	14° 47′	14° 37′	14° 30′	14° 30′	14° 47′	14° 48′
P	(2,1,1) (2,2,0)	29° 05′	29° 00′	28° 46′	29° 03′	39° 03′	29° 00′
	(1,2,1)	23° 45′	23° 40′	23° 30′	23° 50′	23° 40′	23° 40′
	(0,0,2)	22° 26′	—	22° 20′	—	22° 26′	—
	(0,2,1)	20° 00′	—	20° 00′	—	20° 00′	—
	(0,1,1)	16° 40′	16° 40′	16° 25′	16° 42′	16° 40′	16° 38′

* Independent of order.

(From R. A. A. Muzzarelli *et al.*, *Analyst* **94,** 616 (1969).)

complexing polymer having higher capacity or selectivity than the starting materials. Derivatives of cellulose, carrying iminodiacetic, anthranilic and *o*-aminophenylarsinic acid residues, and graft copolymers of cellulose containing acrylamidoxime and acrylhydroxamic acid groups were studied. These polymers collected divalent metal ions, like Cu, Fe, Co, Ni, Zn, Cd, Mn and Ca ions: complex formation was revealed from the comparison of absorption spectra.[30]

Bauman and coworkers worked out a method for diazo-coupling ligands to carboxymethylcellulose in the acid form.[31–32] The CM cellulose was stirred 1 day with *N, N′*-dicyclohexylcarbodiimide (NDCC) and benzidine in methanol solution. The carefully washed product was diazotized in the usual way, and after washing the excess nitrite with sulphamic acid, the diazonium cellulose was stirred in a 5% solution of potassium dithizonate at pH 11. The product was washed free of base with 2 M HCl to yield the active dithizonic acid form. This was washed to neutrality and stored away from air and light which deactivate the chelating polymer so obtained. The same authors claimed the preparation of ligand celluloses in powder form from oxime, cupferron, quinalizarin, 2-thenoyltrifluoroacetone, phenylarsonic acid, and *p*-dimethylaminobenzilidene rhodanine. These ligands coupled readily in 10% aqueous methanol to which a few drops of pyridine were added. The products are purple or red. 250 mg of the dithizone cellulose which was found to contain 0·8 mole of dithizone $mole^{-1}$ of carboxymethyl group, could recover 93% Zn, 70% Cu, 70% Pb from 8 l of 0·12 ppm solutions. Complete recoveries of Cu, Zn, Ag present in 1 l of 1 ppm solution were obtained by passing the solution through a 6×1 cm column of dithizone cellulose. The ligand celluloses were tested in sea-water, where they could collect some transition elements together with sensible amounts of magnesium and calcium. These performances are not exceptional, as they are not higher than those of oxycellulose or carboxymethylcellulose itself, and in any case, the figures reported for recovery of Ni, Sn and Au from 20 l of sea-water cast some doubts on the whole research. The dithizone cellulose is thought to have the following formula:

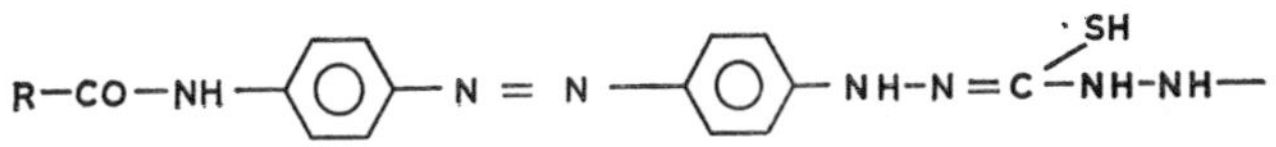

$$R{-}CO{-}NH{-}C_6H_4{-}N{=}N{-}C_6H_4{-}NH{-}N{=}C(SH){-}NH{-}NH{-}$$

The sorption of metal ions from solvent mixtures on carboxymethylcellulose ion-exchangers was studied in more detail by Lasztity and coworkers, further to the researches by the same team. They found that the rate and extent of exchange are strongly increased by adding a basic solvent with a high dielectric constant like water or formamide, to a less basic solvent with a lower dielectric constant, as shown in Fig. 1.16 where data for cobalt are presented.[33]

Also cellulose phosphate has been considered as a chelating polymer.[4] Cellulose phosphate has been shown to have definite advantages over a synthetic polymeric phosphonic acid for the separation of metal ions.[34] By use of eluents containing dilute hydrochloric acid, quantitative column separations of rare earths or aluminium from alkaline earths, and alkaline earths from alkali metal ions were accomplished. Figure 1.17 shows the comparison of two polymers in favour of cellulose phosphate and Fig. 1.18 presents the separation of strontium from europium.

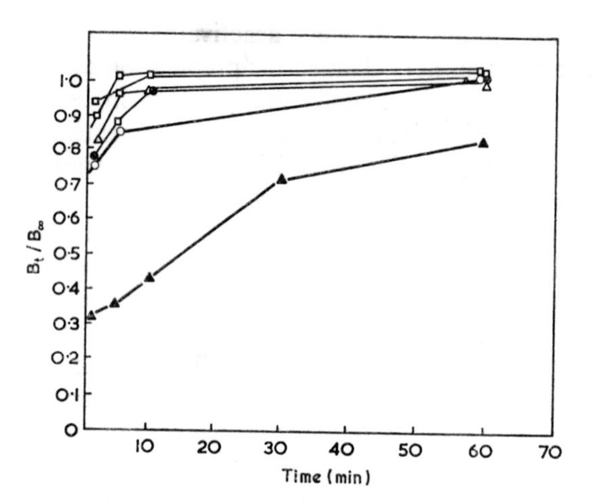

FIG. 1.16. Methanol–water mixtures with mole fractions of ▲ = 0·05, ○ = 0·17, ● = 0·31, ▲ = 0·46, ■ = 0·52, □ = 0·61. B_t = % of total cobalt collected at time t; B_∞ = % of total cobalt collected at equilibrium. (From A. Lasztity *et al.*, *Acta Chim. Acad. Sci. Hung.* **60,** 341 (1969).)

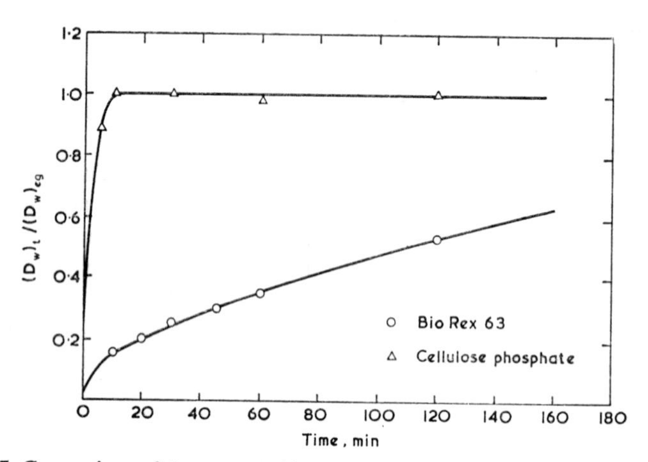

FIG. 1.17. Comparison of the rate at which Bio-Rex 63 and cellulose phosphate attain equilibrium for lanthanum (III) from hydrochloric acid solution. $(D_W)_t$ = weight distribution coefficient at time t. $(D_W)_{eq}$ = weight distribution coefficient at equilibrium. (From D. H. Schmitt *et al.*, *Talanta* **15,** 515 (1968).)

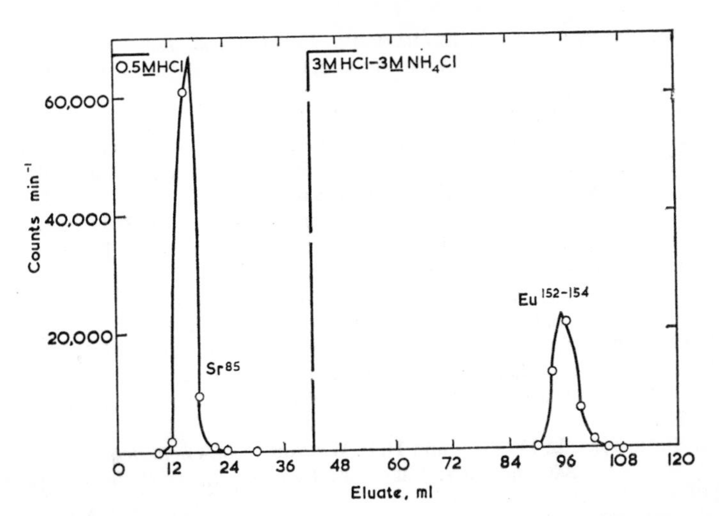

FIG. 1.18. Elution curve for strontium (II) and europium (III) from a cellulose phosphate column. Column: $1 \times 7{\cdot}0$ cm; flow-rate: $1{\cdot}2 \pm 0{\cdot}2$ ml min^{-1}; strontium (II): $1{\cdot}7 \times 10^{-4}$ μmole; europium (III): $6{\cdot}3 \times 10^{-6}$ μmole. (From D. H. Schmitt *et al.*, *Talanta* **15**, 515 (1968).)

REFERENCES

1. R. A. A. MUZZARELLI and G. MARCOTRIGIANO, *Talanta* **14**, 305 (1967).
2. R. A. A. MUZZARELLI and G. MARCOTRIGIANO, *Talanta* **14**, 489 (1967).
3. R. A. A. MUZZARELLI and L. C. BATE, *Talanta* **12**, 823 (1965).
4. R. A. A. MUZZARELLI, Inorganic Chromatography on Columns of Natural and Substituted Celluloses, in *Advances in Chromatography*, Vol. V (R. A. KELLER and J. C. GIDDINGS, Eds.), Marcel Dekker Publ., New York, 1967.
5. C. L. HOFFPAUIR and J. D. GUTHRIE, *Text. Res. J.* **20**, 617 (1950).
6. A. L. LINDSEY, U.S. Patent 891,467 (1962).
7. R. R. BENERITO, B. B. WOODWARD and J. D. GUTHRIE, *Anal. Chem.* **37**, 1693 (1965).
8. G. K. J. GIBSON and D. I. PACKAM, *J. Appl. Chem.* **16**, 50 (1966).
9. D. M. SOIGNET, R. J. BERNI and R. R. BENERITO, *Text. Res. J.* **36**, 978 (1966).
10. E. A. CHAIKINA, L. S. GAL'BRAIKH and Z. A. ROGOVIN, *Cellul. Chem. Technol.* **1**, 625 (1967).
11. D. M. SOIGNET and R. R. BENERITO, *Text. Res. J.* **37**, 1001 (1967).
12. V. O. CIRINO, A. L. BULLOCK and S. P. ROWLAND, *Anal. Chem.* **40**, 396 (1968).
13. E. J. ROBERTS, and S. P. ROWLAND, *Text. Res. J.* **39**, 686 (1969).
14. B. D. DASARE and N. KRISHNASWAMY, *Br. Polym. J.* **1**, 290 (1969).

15. R. M. Noreika and I. I. Zdanavicius, *Cellul. Chem. Technol.* **5**, 117 (1971).
16. R. A. A. Muzzarelli, G. Marcotrigiano, C. S. Liu and A. Frêche, *Anal. Chem.* **39**, 1762 (1967).
17. R. Kuroda and N. Yoshikuni, *Talanta* **18**, 1123 (1971).
18. R. A. A. Muzzarelli, *Talanta* **13**, 809 (1966).
19. R. Kuroda, N. Yoshikuni and K. Kawabuchi, *J. Chromatogr.* **47**, 453 (1970).
20. K. Oguma and R. Kuroda, *J. Chromatogr.* **61**, 307 (1971).
21. R. Kuroda, T. Kiriyama and K. Ishida, *Anal. Chim. Acta* **40**, 305 (1968).
22. G. Bagliano, L. Ossicini and M. Lederer, *J. Chromatogr.* **21**, 471 (1966).
23. K. Ishida and R. Kuroda, *Anal. Chem.* **39**, 212 (1967).
24. S. Kaufman and L. S. Keyes, *Anal. Chem.* **36**, 1777 (1964).
25. L. C. Lock and E. C. Martin, *J. Chromatogr.* **7**, 120 (1962).
26. E. G. Davidova and V. V. Rachinsky, *Prikl. Biokhim. Mikrobiol.* **3**, 341 (1967).
27. B. A. Zabin, U.S. Patent 3,423,396 (1969).
28. R. A. A. Muzzarelli, A. Ferrero Martelli and O. Tubertini, *Analyst* **94**, 616 (1969).
29. K. Shimomura, L. Dickson and H. F. Walton, *Anal. Chim. Acta* **37**, 102 (1967).
30. N. V. Tolmachev and L. V. Mirosnik, *Vyzokomol. Soedim.* Ser. A, **10**, 1811 (1968).
31. A. J. Bauman, H. H. Weetall and N. Weliky, *Anal. Chem.* **39**, 932 (1967).
32. A. J. Bauman, N. Weliky and H. H. Weetall, U.S. Patent 3,484,390 (1969).
33. A. Lásztity and M. Osy, *Acta Chim. Acad. Sci. Hung.* **60**, 341 (1969).
34. D. H. Schmitt and J. S. Fritz, *Talanta* **15**, 515 (1968).

CHAPTER 2

ALGINIC ACID

PRESENTATION

Alginic acid is a polyuronide found in brown seaweeds whose chemical formula is reported below:

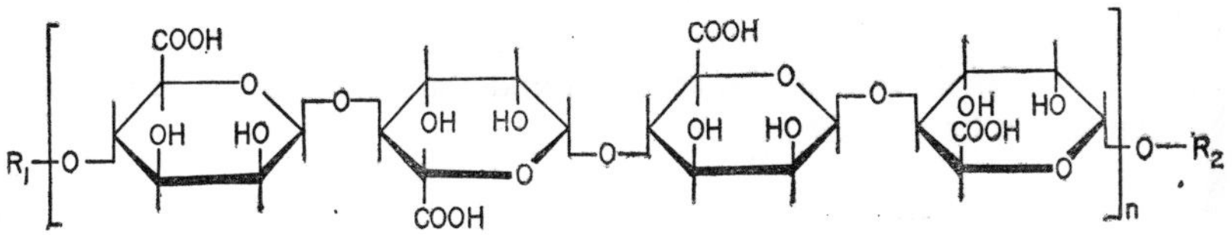

It is commonly obtained from *Macrocystis pyrifera*, *Laminaria* species and *Eklonia* species. The alginic acid of these species varies in the proportion of 14 to 40% of the dry solids of these seaweeds. The alginate in seaweeds behaves as a base exchange material, and from the knowledge of the composition of the salts in the seaweed, the conclusion was reached that alginic acid is present as a mixed salt of cations, mostly calcium, able to render it insoluble. Through divalent ions, alginic acid may also be combined to other substances.[1–3]

Isolation of alginic acid presents the obvious problem of avoiding degradation. The seaweed is reduced to powder, and using base exchange reactions, the alginic acid is alternatively made soluble and insoluble, so as to separate it from other components.[4] The preliminary treatment involves hot water and lime water giving a solution of mannitol, laminarin, fucoidan and salts. Then, a dilute acid treatment eliminates acid soluble substances, while converting the alginate to free alginic acid. By stirring the residue with sodium carbonate solutions, alginic acid is dissolved as soluble sodium alginate. Precipitation with alcohol has been proposed to free the alginic acid from colouring matter. However, to perform a complete purification, several precipitations are required following different techniques which are described in patents.[5, 6]

The research on alginic acid and its derivatives has produced many publications and patents. According to Steiner and McNeely,[3] the patents granted before 1959 were 740, and in the following decade about 100 more patents were also obtained. Therefore the description of the production of alginic acid and its derivatives is beyond the scope of this book; for the purpose of presenting an example an industrial process is summarized below.[4]

Ascophyllum nodosum is treated with sulphuric acid and extracted with soda. After filtering to remove insoluble residues, potassium chloride is added to the alginate to a concentration of 0·1 N and sulphuric acid is added to a pH = 2·1. This causes precipitation of the alginic acid. Filtration yields fibrous alginic acid. The process is conducted at 10–15 °C It is important that the amount of sulphuric acid added be just equivalent to the alginic acid in the seaweed and that the potassium chloride concentration be $\geq$ 0·07 N.

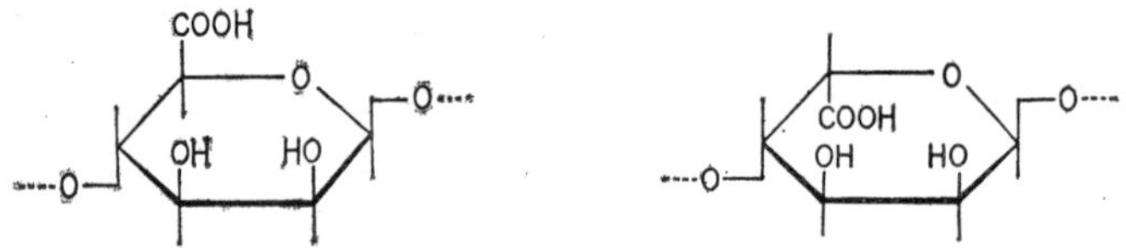

Mannuronic and guluronic acids are the two constituents of alginic acid: this information was reached after Nelson and Cretcher established that the uronic acid content of alginic acid is 100% based on a decarboxylation reaction, and that mannuronic acid was a constituent of the polymer.

Following studies of methylation, hydrolysis and oxidation by many authors,[6–16] it was demonstrated that variable amounts of *L*-guluronic acid were also present in alginic acid samples from seventeen different genera of seaweeds. Evidence that the guluronic acid residues were also linked through C–1 and C–4 was supported by the identification of the two substituted pentaric acids isolated after alkaline degradation of alginic acid.

Apparently, alginic acid is a heteropolymer including both the uronic acids, but there is no evidence against the hypothesis of two different polymers, one made up solely of mannuronic acid, and the other of guluronic acid.

In connection with this type of investigation, a certain amount of information was obtained on the resistance to hydrolysis of alginic acid, which, like all polysaccharides with a high uronic acid content is very difficult to hydrolyse. By hydrolysing alginic acid with 1 M oxalic acid at 100 °C for 10 hr, Haug[17] found that 28% of the alginate passed in solution, and that there was a rapid depolymerization. The redissolved and reprecipitated residue submitted to a similar treatment was dissolved to the extent of 19% with no appreciable depolymerization, in accordance with the removal of hydrolysable material mainly from chain ends.

The hydrolysate contained monomeric guluronic and mannuronic acids, and unidentified diuronides. On the basis of chromatographic separations carried out on the hydrolysates, it was suggested that the molecule is made of chains of mannuronic acid or guluronic acid twenty to thirty units long, alternating to sections where the two uronic acids are about in the same proportion. The homogeneous sections seem to be protected from random hydrolysis by their crystalline character.

MACROSTRUCTURE

In view of the preceding discussion, one can imagine that instrumental evidence for a clear macrostructure would be hardly attainable. Several authors[18–25] studied alginic acid fibres prepared by extrusion and stretched when still in the wet state: these fibres show a clear X-ray diffraction pattern similar to cellulose. Astbury calculated that the distance along the fibre axis was 8·7 Å instead of 10·3 Å for cellulose. Frei and Preston suggested that the recorded X-ray pattern was due to polyguluronic acid, but as it is unlikely that pure polyguluronic acid was isolated, there is much difficulty in interpreting the results.

This confusion is associated with the fact that in mixtures, crystallization of polyguluronic acid tends to suppress that of polymannuronic acid. The interplanar spacings of these two acids are listed in Table 2.1 together with those of the cellulose typical of brown algae.

On the other hand, calcium alginate does not give a sharp X-ray diffraction spectra, and the moisture retained by the polymer further complicates the patterns.

No X-ray diffraction record is available for other elements fixed on alginic acid, and in any case the study in this direction would possibly not be fruitful because of the complicate macrostructure of the polymer itself. Metal ions are also expected to make the structure less regular, as in the case of calcium.

TABLE 2.1. X-RAY SPACINGS OF POWDERS OF THE TWO MAJOR COMPONENTS IN ALGINIC ACID IN AN ATMOSPHERE OF 98 PER CENT RELATIVE HUMIDITY AND THE PROMINENT SPACINGS OF CELLULOSE, ALL FROM *Himanthalia elongata*

Polymannuronic acid*	Polyguluronic acid*	Cellulose
6·53 *W*	6·65 *VS*	
		6·10 *M*
5·66 *S*		
	5·26 *M*	5·25 *M*
5·07 *VW*		
4·32 *VS*	4·27 *MW*	4·32 *W*
	3·94 *S*	3·90 *S*
3·81 *M*	3·84 *S*	
	3·61 *M*	
3·56 *M*		
3·33 *VW*		
	3·09 *M*	
2·96 *W*		
2·72 *W*	2·75 *W*	
	2·64 *W*	
2·56 *VW*		2·58 *M*
2·47 *M*		
2·36 *W*	2·39 *VW*	
	2·16 *VW*	

VS, very strong, *S*, strong, *M*, medium, *MW*, medium weak, *W*, weak, *VW*, very weak, rings on diagram.

* Some of the spacings vary with water content.

(From E. Frei and R. D. Preston, *Nature, Lond.* **196,** 133 (1962).)

Occasionally, an infrared spectrum was reported[26] for alginic acid and cobalt alginate (Fig. 2.1) where the strong band at 1735 cm^{-1} was assigned to the free carboxyl group of alginic acid, while the band at 1600 cm^{-1} was assigned to the salified carboxyl group. In any case, this instrumental evidence of salification does not add new light to the existing knowledge about interaction of metal ions with the polymer, as the salification reactions are well known from precipitation studies, and the chelating ability of the polymer is not evident from the hydroxyl group region of the infrared spectrum.

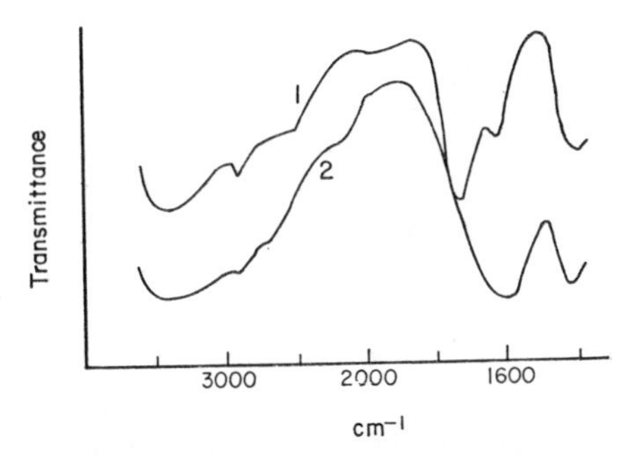

FIG. 2.1. I.R. spectra of (1) alginic acid and (2) Co alginate. (From D. Cozzi *et al.*, *J. Chromatogr.* **35**, 396 (1968).)

The molecular weight of alginic acid was determined by measurement of the viscosity of the alginate solutions, 0·1 N sodium chloride having been used to depress the polyelectrolyte effect.[27] Cook and Smith critically examined the various suggested relations between molecular weight and intrinsic viscosity.[28] A degree of polymerization of 1000 for the longest polymer chains was indicated, even though the molecular weight distribution is not known.

CHARACTERISTICS OF ALGINIC ACID

The dissociation constant measurements done before 1961 are not reliable because of lack of information about the presence of varying amounts of guluronic acid in the polymer. Those alginates having the highest proportion of guluronic acid exhibit the highest pK values. In fact mannuronic acid has a pK value of 3·38 and guluronic acid 3·65. These figures are for 0·1 N sodium chloride solutions where alginic acid from *L. hyperborea* gave 3·74 and alginic acid from *L. digitata* gave 3·42.[29, 30]

As each chain unit, independently of its stereoconfiguration, has a constant stoichiometry, the equivalent weight for the free acid should be 176. Values close to the theoretical one have been reported for samples well dried over phosphorous pentoxide at 60 °C for 24 hr but samples currently dried at 105 °C generally give higher figures around 200. In several cases the recorded value is 216, due to some water being retained by the polymer.[31]

Of course it is easy to titrate such an acid, and this is done currently for determination of the alginic acid concentration in the solutions obtained after seaweed treatment. For instance, 100 mg of alginate at the concentration of 0·1 can be precipitated by calcium chloride in excess, by washing free from calcium with hydrochloric acid, and from chloride with water, it can then be titrated with sodium hydroxide with phenolphthalein as indicator.

An indirect titration was also proposed, which has the advantage of avoiding delay in titration due to increased viscosity of alginate solution. Calcium acetate is added to alginic acid, and the liberated acetic acid is titrated. The resulting solution can be further used to determine viscosity for the polymer evaluation, after addition of sodium salts.[32]

When alginate is heated in the absence of acid, carbon dioxide is quantitatively released, so that it can be used as a measure of the uronide present in seaweed. On heating alginate with concentrated hydrochloric acid carbon dioxide is evolved, and Jensen demonstrated that the method can be applied directly to seaweeds for the determination of alginic acid.[33, 34]

Colorimetric methods have also been developed. Orcinol in hydrochloric acid has been proposed by Brown and Hayes.[35] Absorption is enhanced by borate. The results, however, may depend on the proportion of the two uronic acids present in the sample. An automated procedure based on the colorimetric measurements obtained with carbazole in concentrated sulphuric acid has also been worked out.[36]

The proportion of guluronic acid and mannuronic acid in samples of alginate can be known by performing column chromatography on anion exchange resins (Dowex 1). This procedure takes into account the fact that both uronic acids form lactones and free acids after hydrolysis.[29] Therefore alginate is pretreated with 80% sulphuric acid at 0 °C; the mixture is left to stand at room temperature for 18 hr. The mixture is diluted to 2 N on cooling with ice, and then heated in a sealed tube on boiling water for 5 hr. The pH is then adjusted to 8·0 with calcium carbonate, and the acids are separated on the column preconditioned in the acetate form. For a correct estimation of the ratio of mannuronic to guluronic acid it should be multiplied by 0·66 to take into account the greater degradation of guluronic acid during the above procedure.

Other estimates[10] involve paper chromatography in pyridine+ethyl acetate+water, and ionoforetic separations in 0·01 M sodium tetraborate+0·005 M calcium chloride with a current of 0·5 mA cm^{-1}: the mobilities of guluronic, mannuronic, galacturonic and glucuronic acids are respectively 0·74, 0·88, 1·03 and 1·26 whereas glucose made 1·00.[30]

AFFINITY OF DIVALENT METAL IONS TO ALGINATES

The amount of divalent ions necessary to obtain precipitation of alginates increases in the order Pb, Cu < Ca < Co, Ni, Zn < Mn.[30] The ion exchange properties of alginates depend on the chemical composition of

TABLE 2.2. pH OF ALGINATE SOLUTIONS CONTAINING METAL SALTS

Metal salt	mequiv. salt / mequiv. alg	*L. digitata*			*L. hyperborea* stipe		
		0	0·83	1·66	0	0·83	1·66
$Pb(NO_3)_2$		3·90	2·60	2·40	4·48	2·60	2·38
$Cu(NO_3)_2$			2·85	2·75		2·95	2·65
$Cd(NO_3)_2$			2·94	2·77		3·13	2·96
$Ba(NO_3)_2$			3·00	2·80		3·05	2·85
$Sr(NO_3)_2$			3·05	2·95		3·10	2·92
$Ca(NO_3)_2$			3·21	2·98		3·25	3·02
$Co(NO_3)_2$			3·24	3·11		3·41	3·31
$Ni(NO_3)_2$			3·32	3·15		3·51	3·38
$ZnSO_4$			3·32	3·22		3·49	3·40
$MnSO_4$			3·32	3·31		3·74	3·72
$MgSO_4$			3·39	3·36		3·80	3·70

(From A. Haug, *Acta Chem. Scand.* **15**, 1794 (1961).)

TABLE 2.3. URONIC ACID COMPOSITION OF ALGINATES

Source of alginate	$k_{Cd\text{-}Sr}$	Mannuronic acid / Guluronic acid
L. digitata, Krákvágoy 3/7	1·17	2·10
L. digitata, Espevaer 8/1	0·98	1·75
L. hyperborea stipes, Reine 28/6	0·70	1·10
L. hyperborea stipes, Hustad 26/2	0·54	0·56
Ascophyllum nodosum, Vaere 2/3	1·19	2·45
Commercial samples:		
Protanal LF *(L. digitata)*	1·22	2·17
Manucol SS/LD *(L. hyperborea* stipes)	0·58	0·70
Manucol SA/LM *(A. nodosum)*	0·95	1·65

(From A. Haug, *Acta Chem. Scand.* **15**, 1794 (1961).)

the alginate. Alginates rich in mannuronic acid such as those from *Laminaria digitata* have a lower affinity to calcium in a sodium–calcium ion exchange reaction than alginates rich in guluronic acid units, such as the alginate from *Laminaria hyperborea.*

When a metal salt solution is added to a half neutralized sample of alginic acid, the pH decreases and this is an indication of the affinity between the metal ion and the alginate. In Table 2.2 results are presented for the two alginates mentioned above. The uronic acid composition of the two alginates used is reported in Table 2.3. While the *L. digitata* alginate follows the ionotropic series of Thiele based on the capillary diameter of gel formed by allowing divalent metals to diffuse into alginate solutions, the *L. hyperborea* alginate does not. The ionotropic series of Thiele is as follows: Pb$<$Cu$<$Cd$<$Ba$<$Sr$<$Ca$<$Zn, Co, Ni. In Table 2.3 the selectivity coefficients for the cadmium–strontium reaction of exchange are also reported.

These coefficients have also been calculated for calcium-strontium ion exchange reactions.[38–39] The formation constants of the polyuronic acids were found by the ion exchange resin procedure developed by Schubert, which consider free metal ions in solutions with no polymerization or hydrolysis. The formation constant of a complex is given by:

$$K_f = \frac{(K_d^{\circ}/K_d) - 1}{A^n}$$

where K_d are the distribution coefficients of the metal ion between resin and solution in the absence (°) and presence of alginate (ligand). A is the molar concentration of alginate assuming an equivalent weight of 198,

TABLE 2.4. FORMATION CONSTANTS OF THE CALCIUM AND STRONTIUM COMPLEXES OF A SERIES OF ALGINATE DERIVATIVES OF VARYING GULURONIC ACID CONTENTS AND DEGREES OF POLYMERIZATION (D.P.)

Percentage guluronic acid	D.P.	K_fCa	K_fSr	$\frac{K_f\,\text{Sr}}{K_f\,\text{Ca}}$
96	13	104	449	4·3
93	8	124	443	3·6
86	226	109	429	3·9
72	75	110	353	3·2
61	57	119	317	2·7
41	440	143	208	1·5
30	74	80	163	2·0

(From J. T. Triffitt, *Nature, Lond.* **217**, 457 (1968).)

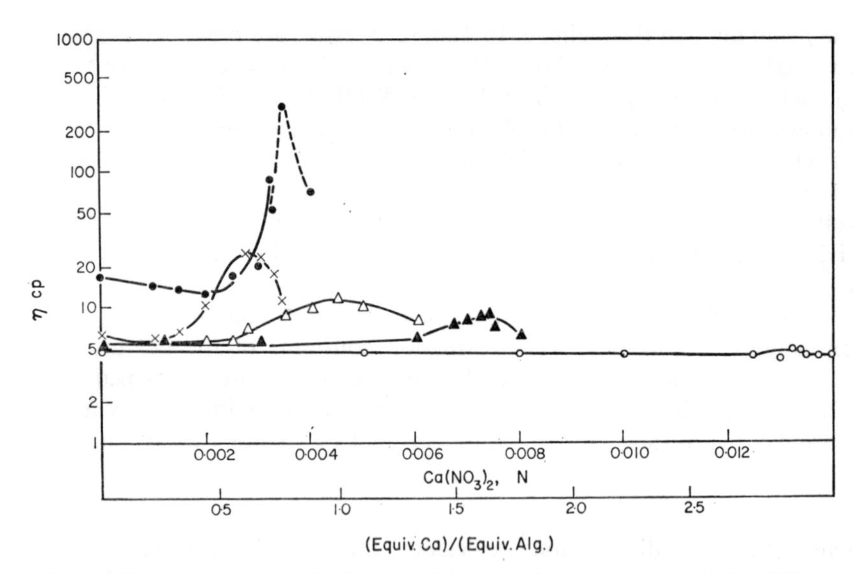

FIG. 2.2. Viscosity of 0·1% (0·0045 N) solutions of sodium alginate containing different amounts of sodium nitrate after addition of various amounts of calcium nitrate. Alginate from *L. digitata*, Tarva 3/7/57, $[\eta] = 26{\cdot}0$ (100 ml/g). $NaNO_2$: ● = 0 N; × = 0·05 N; △ = 0·1 N; ▲ = 0·2 N; ○ = 0·4 N. —— = Pipette No. 1; - - - - = Pipette No. 2. (From O. Smidsrod *et al.*, *Acta Chem. Scand.* **19**, 329 (1964).)

and n is the number of moles of ligand relative to the metal in the associated molecule. The sodium uronides were equilibrated with a cation exchange resin (Dowex 50) and by radiochemical techniques the results in Table 2.4 were obtained. From these results one can establish a correlation between guluronic acid content and strontium alginate formation constants, while the data for calcium is more uniform. The relative binding of strontium compared with calcium tends to increase with guluronic acid content.[39]

When calcium ions are added in increasing amounts to an alginate solution, the viscosity of the solution increases to a maximum value. Further addition of calcium leads to precipitation resulting in a decrease in viscosity. In Fig. 2.2 results are presented for solutions containing various amounts of sodium nitrate. The amount of sodium ions present in solution, influences the viscosity and the amount of calcium ions necessary to reach the viscosity maximum. In fact at low concentrations of sodium ions the viscosity maximum is displaced towards lower calcium ion concentrations.

When calcium nitrate is added to a solution of sodium alginate, a cation exchange takes place, following the general reaction where Me^{++} is a divalent metal ion:

$$\underset{\text{gel}}{Me(Alg)_2} + \underset{\text{liquid}}{2\,Na^+} = \underset{\text{gel}}{2\,NaAlg} + \underset{\text{liquid}}{Me^{++}}.$$

When the concentrations in the gel are expressed as equivalent fractions and those in the liquid as normalities the equilibrium constant is called

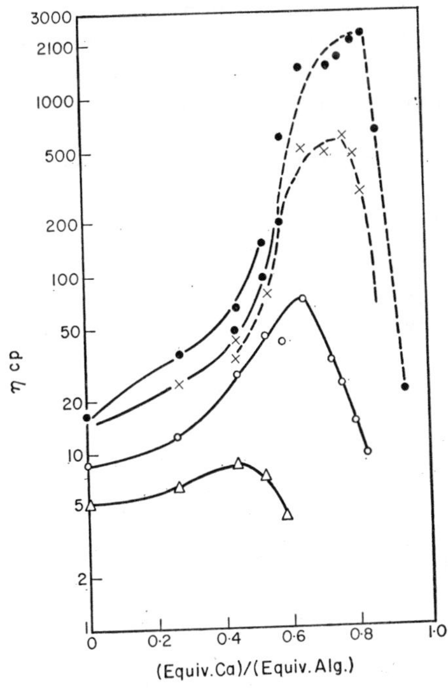

FIG. 2.3. Viscosity of 0·1% solutions of mixed sodium–calcium alginates. Preparations of sodium–calcium alginates dissolved in water and mixed with sodium nitrate solutions of different strengths.

Alginate from *L. digitata*, Tarva 3/7/57, $[\eta] = 26{\cdot}0$.

$NaNO_3$: ● = 0 N; × = 0·001 N; ○ = 0·005 N; △ = 0·05 N; —— = Pipette No. 1; – – – – = Pipette No. 2.

(From O. Smidsrod *et al.*, *Acta Chem. Scand.* **19**, 329 (1965).)

selectivity coefficient:

$$\frac{[Me^{++}\ gel]\ [Na^{+}\ liq]^2}{[Na^{+}\ gel]^2\ [Me^{++}\ liq]} = K.$$

After addition, the equivalents of free cations consisting both of sodium and calcium ions will be equal to those of calcium added. The ionic strength increases and this causes a slight decrease of viscosity for low calcium concentrations in the samples where no sodium nitrate is present.

In order to investigate the effect of small quantities of free cations, preparations of mixed sodium–calcium alginate were dissolved in water and mixed with solutions containing the desired amount of sodium nitrate.[40] The viscosity as a function of the calcium content is presented in Fig. 2.3. A marked decrease of the viscosity and an important displacement towards lower values of calcium concentrations are observed when larger amounts of sodium are present.

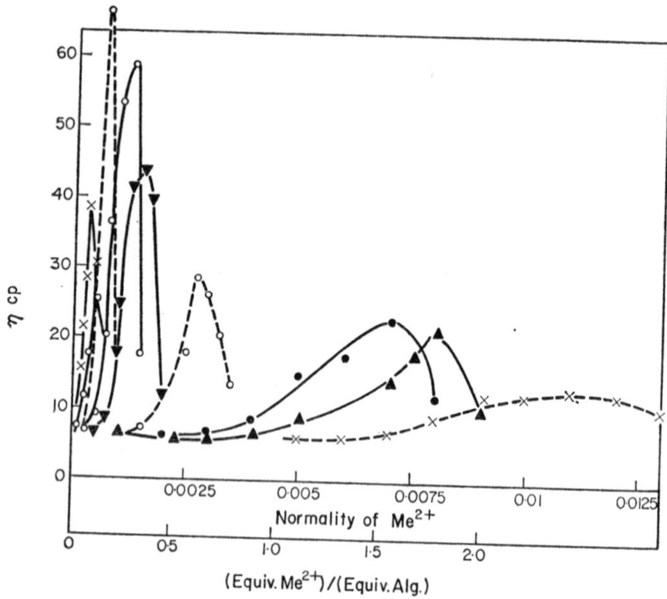

FIG. 2.4. Viscosity of 0·1% (0·0045 N) alginate solutions containing divalent metals. Alginate from *L. digitata*. Sodium nitrate concentration 0·05 N.

×—× = Ba; △—△ = Pb; ○---○ = Cu; ○——○ = Sr; ▼——▼ = Cd; ○---○ = Ca; ●——● = Zn; ▲——▲ = Ni; ×---× = Co.

(From A. Haug *et al.*, *Acta Chem. Scand.* **19**, 341 (1965).)

Similar experiments with some divalent metal ions were also reported.[38] With exception of magnesium, manganous and ferrous ions, the divalent metal ions tested gave a viscosity maximum at a well defined concentration of metal ion. This concentration (Me_{max}) is very different for each metal ion as reported in Figs. 2.4 and 2.5, and is characteristic of the precipitation effect of the metal ion upon alginic acid. It is possible

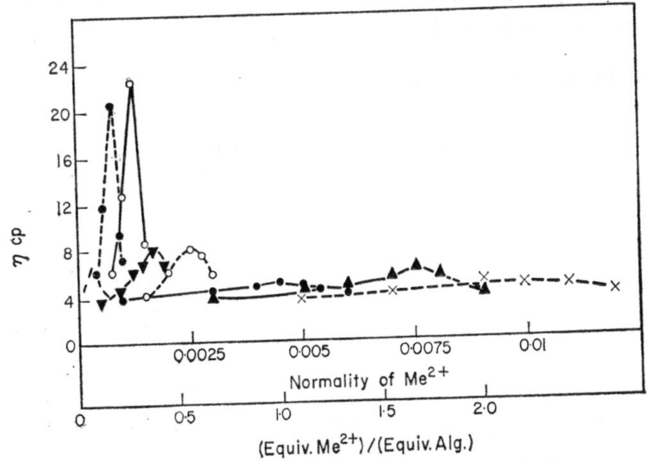

FIG. 2.5. Viscosity of a 0·1% alginate solution containing divalent metals. Alginate from *L. hyperborea* stipes. Sodium nitrate concentration 0·05 N. Symbols are the same as in Fig. 2.4.
(From A. Haug *et al.*, *Acta Chem. Scand.* **19**, 341 (1965).)

to see that the results reported in these figures do not strictly correspond to the affinity series of alginates, reported above in Table 2.1.

When the selectivity coefficient of the ion exchange is known, it is possible to calculate the amount of metal ion bound to alginic acid ($Me_{g\ max}$). Some selectivity coefficients are given in Table 2.5, while in Table 2.6 are listed the amounts of metal which must interact with the alginic acid to produce gel formation. As $Me_{g\ max}$ differs for each metal, the precipitation effect of a certain divalent metal is determined by:

1. the affinity of the metal ion for alginic acid which determines the amount of metal bound to the alginate in a solution containing a certain concentration of metal ions, and

2. the gel-forming ability of the metal ion, which determines the amount of metal which must be bound to the alginate to precipitate it.

The sodium ions were found to have two opposite effects:

(i) they displace the ion exchange equilibrium in the direction of lower values of Ca^{++}_{gel}, and

(ii) they produce a salting out effect, because, as said above, they affect the ionic strength of the solution.

This can be seen in Table 2.6.

TABLE 2.5. SELECTIVITY COEFFICIENTS, k, FOR DIFFERENT ION EXCHANGE REACTIONS

$$k = \frac{Me_g \cdot [Na_l]^2}{[Me_l] \cdot Na_g^2}$$

Metal ions	*L. digitata*, Tarva 3/7/57 M/G = 1·60	*L. hyperborea* stipes, Hustad 12/2/59 M/G = 0·45
Cu^{2+}–Na^+	230	340
Ba^{2+}–Na^+	21	52
Ca^{2+}–Na^+	7·5	20
Co^{2+}–Na^+	3·5	4

(From A. Haug *et al.*, *Acta Chem. Scand.* **19**, 341 (1965).)

For metal ions with high affinity for alginate, like copper, the salting out effect is the most important one, and therefore less copper is needed to form a gel, when sodium ion concentration is higher. For metals with low affinity, like cobalt, the ion exchange effect predominates, and more cobalt ions are necessary for obtaining gel formation at higher sodium concentrations.

The selectivity coefficients given in Table 2.5 show that guluronic rich alginates have a higher affinity for divalent metal ions than mannuronic rich alginates; this different behaviour is particularly interesting when manganous ions are considered, as an alginate fractionation can be carried out with manganous ions. Manganous ions produce a flocculent pre-

TABLE 2.6. VISCOSITY MAXIMUM FOR DIFFERENT DIVALENT METALS. CONCENTRATION OF ALGINATE: 0·1%, 0·0045 equiv/l.

	L. hyperborea stipes, Hustad 12/2/59 [η] = 16·4, M/G = 0·45 0·05 N sodium nitrate.		*L. digitata*, Tarva 3/7/57 [η] = 26, M/G = 1·60 0·05 N sodium nitrate.		0·2 N sodium nitrate.	
Metal	Me_{max}	$Me_{g\ max}$	Me_{max}	$Me_{g\ max}$	Me_{max}	$Me_{g\ max}$
Ba	0·067	0·066	0·067	0·065	0·033	0·025
Pb	0·111		0·111		0·045	
Cu	0·178	0·178	0·155	0·155	0·13	0·12
Sr	0·278		0·278		0·33	
Cd	0·378		0·333		0·67	
Ca	0·555	0·49	0·61	0·47	1·65	0·39
Zn	1·00		1·56		2·45	
Ni	1·66		1·78		3·33	
Co	2·22	0·72	2·45	0·71	5·55	0·50

(From A. Haug *et al.*, *Acta Chem. Scand.* **19,** 341 (1965).)

cipitate while the other divalent metals form a gel which takes the total volume of the mixture, unless high ionic strength solutions are used. At high ionic strength fractionation is also possible with calcium ions.[41]

If an alginate gel is formed by the diffusion of a divalent metal ion into an alginate solution, the gel is birefringent, because during the precipitation process orientation of the polymer chains occurs.[42] The gel formed has a smaller volume than the original alginate solution and the greater the shrinkage, the greater is the birefringence. Under the same conditions the shrinkage and the birefringence follow the ionotropic series, in the order of maximum shrinkage and birefringence for lead and minimum for manganese.

High orientation calcium alginate gels can be obtained by allowing calcium chloride to diffuse through a membrane into a flowing stream of sodium alginate solution. Sterling has published data on their birefringence and X-ray diffraction spectra.[22]

From the preceding discussion it is clear that solubility of alginates and alginic acid is affected by many factors. Alginic acid is precipitated by the addition of strong acids to an alginate solution. With alginates obtained from seaweeds the acid is precipitated over a pH range which depends on the species from which it derives, on the degree of polymerization and on the ionic strength.[43] A soluble alginic acid from *A. nodosum* was also reported.[44] Percival and McDowell suggest that the difference in solubility is a result of a difference in the sequence of mannuronic and guluronic acid residues.[1]

Alginates of alkali metals and ammonia and a few organic bases are also soluble in water. Unlike its group homologues, magnesium yields a water soluble alginate. Alginates have no clearly defined solubility limits, and upon increasing the concentration of alginate the solutions become more and more viscous until a paste or a plastic solid results.

HEALTH PHYSICS EXPERIMENTS

As alginate was found to be a most effective agent for inhibiting intestinal strontium uptake, without interfering appreciably with the metabolism of calcium and the alkali metal ions, it was proposed for prevention and therapy of strontium radiocontamination.[45]

In vivo experiments

Animal experiments reported by several scientists demonstrated the selective inhibition of the gastrointestinal absorption of strontium. Skoryna[46] observed a five-fold reduction of Sr absorption in rats by sodium alginate: the observed ratios being 0·32 in control animals down to 0·15 when alginate together with ^{89}Sr and ^{45}Ca were administered with the diet. A six-fold reduction was recorded when sodium alginate was added to ^{85}Sr contaminated milk supplied to young swine. The observed ratio went down from 0·27 in the controls to 0·07 in the alginate treated animals. Harrison[47] obtained in rats a 4·5-fold decrease of skeletal burden whereas the observed ratio went down from 0·23 in the controls to 0·07 in the alginate treated animals. With alginate rich in guluronic acid residues the reduction was six-fold.

Patrick[48] obtained a four–six-fold decrease in the uptake of ^{85}Sr in rats with no interference in the ^{45}Ca uptake. Kostial[49] reported a 6·5-fold decrease of skeletal ^{85}Sr by giving a diet supplemented with sodium alginate and calcium phosphate. Incidentally it should be observed that in this investigation as well as in some other, both ^{47}Ca and ^{85}Sr radiotracers are administered carrier-free but in practice ^{47}Ca is representative of itself plus 1 g of calcium contained in 100 g of diet, while ^{85}Sr is representative of itself only.

By starting alginate administration at the same time of radiocontamination a 10-fold reduction of ^{85}Sr was obtained by Stara.[50]

As a dry substance, alginate *in vivo* is not effective, following experiments on chicks and pigs. The same negative results are verified in rats after giving them 12 mg sodium alginate and 18 mg strontium by stomach tube.[51, 52]

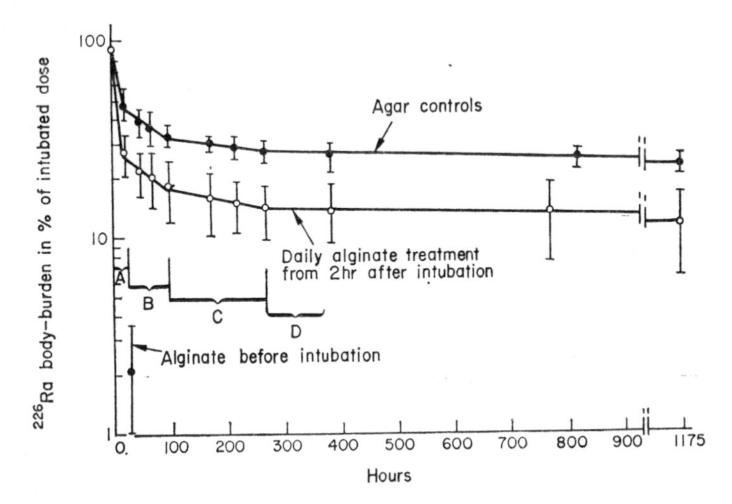

FIG. 2.6. Body-burden of ^{226}Ra in mice: influence of daily therapeutic alginate treatment. The isotopes were administered by gastric tube, alginate administration started 2 hr after intubation and was repeated daily (about 20 mg/day). *In vivo* measurements on Ge(Li) detector, 4–6 mice per point, limits = 95% confidence interval. Semi-log scale: note sequence of exponential decreases (B, C, D) of the ^{226}Ra body-burden. (From O. Van Der Borght *et al.*, *Health Phys.* **21**, 181 (1971).)

Van Der Borght and coworkers[45] studied the intestinal absorption and body retention of ^{226}Ra and ^{47}Ca in mice and measured the effects of sodium alginate *in vivo* on the two radioisotopes by gamma-ray spectrometry with a Ge(Li) detector (Fig. 2.6). For this study, agar was administered to control mice. To compare the influence of different alginates, mice were allowed to eat agar gels, alginate gels, bread made with addition of agar and bread with alginates after a fasting period of 12 hr.

TABLE 2.7. LOWERING OF ^{226}Ra AND ^{47}Ca IN FEMUR OF MICE, BY A SINGLE MEAL OF DIFFERENT KINDS OF ALGINATES BEFORE ISOTOPE ADMINISTRATION

Alginate	Reduction factor for ^{226}Ra in femur	OR femur/intubated dose
1. Alginate + agar + cheese gels		
Agar controls	1	0·6 ± 0·2
Meadows SS/LD/2	70	0·02 ± 0·01
Kelgin XL	9	0·05 ± 0·02
Smith & Sons	7	0·10 ± 0·05
Fluka "purum"	20	0·04 ± 0·04
2. Alginated bread		
Agar-bread	1·2	0·3 ± 0·2
SS/LD/2-bread	135	0·010 ± 0·006
Fluka-bread	110	0·010 ± 0·006

Five mice per group, access to diet 2 hr before intubation of isotopes.
Isotopes given by gastric tube, femurs dissected 6 days later.
Statistical limits = 95% P confidence intervals.

$$\text{OR femur/intubated dose} = \frac{\%\ ^{226}\text{Ra of dose in femur}}{\%\ ^{47}\text{Ca of dose in femur}}.$$

Reduction factor = Ra in control/Ra in test.

(From O. Van Der Borght *et al.*, *Health Phys.* **21,** 181 (1971).)

After a 2-hr eating period the quantities of the ingested food were weighed and the ^{226}Ra with ^{47}Ca doses were intubated. Two hours later the mice had access to their standard diet, and they were sacrificed 6 days after intubation: results are reported in Table 2.7.

The reduction factors are particularly high with alginate breads. Similar breads prepared with the equivalent addition of agar did not appreciably lower the ^{226}Ra incorporation.

When fasting mice were first intubated with ^{226}Ra and a therapeutic alginate treatment started 2 hr later, the per cent body burden is reduced in four periods and can be considered as a composite exponential function. After the faecal excretion of the unabsorbed part of the dose, (period A), period B mainly corresponds to the elimination of ^{226}Ra by desquamation of the intestinal epithelium where some ^{226}Ra is absorbed at the time of administration. *In vivo* experiments on rats with ^{14}C labelled alginate confirmed that far less than 15% of the administered alginate was assimilated by the animals.[53] In practice the incorporation of alginate into bread seems to be more favourable for the mixing with the $^{226}RaCl_2$ solution and for the protection of the epithelium with an alginate layer. The influence of alginate is not due to a shortening of the intestinal transit time, as agar with similar viscosity was administered to the control animals. The influence of alginate can be attributed to the binding of Sr and Ra, which is favoured by respect to the binding of Ca.

In piglets fed with milk obtained from a cow to which ^{134}Cs and ^{85}Sr were injected the body burdens obtained with and without alginate additions were compared.[54] The biological decay curve is a simple exponential function in the case of caesium. The caesium body burden is not appreciably lowered by the alginate administration; at the tenth day, for example, the radioactivity in pigs fed without alginate amounted to 18·5% ±5% of the administered dose whereas with alginate the radioactivity amounted to 10·2% ±6·4%; this slight difference disappeared after 30 days. For ^{134}Cs the biological half-life as calculated from the theoretical regressions, is 12·2 days with alginate, and 12·3 days without alginate; the results for ^{85}Sr are of course much more divergent. A difficulty encountered in this study was the very high viscosity of liquids when as little as 5% alginate is added to milk: the hungry swine had difficulty to ingest more than 500 ml of the stiff milk paste.

A summary of the results obtained for strontium has recently been published.[45]

In vitro experiments

With rat duodenal slices Patrick[55] obtained a 30% reduction of the ^{85}Sr fixation while Ca was reduced by 10%. He also observed *in vitro* the importance of the guluronic acid content of the alginate for the inhibition of the Sr uptake by duodenal slices.[48] With ileal segments of rats, transfer of Sr and Ca ions through the intestinal wall was reduced by 41% and 15% respectively by alginic acid.[56]

Experiments in man

The inhibition of the intestinal uptake of Sr by sodium alginate was also confirmed in man. Hesp[57] found a nine-fold reduction of the Sr uptake in adult man. Sutton[58] also reported a three–five-fold reduction in plasma ^{87}Sr levels when the radioisotope was administered *per os* together with sodium alginate. In this investigation the trend of higher guluronic acid content, higher reduction of plasma strontium level was confirmed. The Sr–Ca selectivity coefficients for alginates of varying uronic acid composition are presented in Fig. 2.7.

A twentyfour-fold reduction of Sr uptake was observed after administering Ca and Sr isotopes with alginate to patients;[59] the ratio went from 0·6 in the control patients down to 0·09 in the alginate treated patients. It is interesting to note some data reported on transition metals; in the two patients who received 10 g sodium alginate per day for 7 days the urinary excretion of magnesium, iron and copper was decreased, while the faecal excretion of strontium, calcium, magnesium, iron, copper and zinc was increased.[59] No change in the excretion of sodium, potassium, magnesium and phosphorous was observed by other authors.[60] All these data are in agreement with the chelating and ion-exchange role of alginic acid. Some data are in Table 2.8.

TABLE 2.8. UPTAKE OF STRONTIUM-85 AND CALCIUM-45 BY DUODENAL SLICES IN THE PRESENCE OF POLYURONIDES AND RELATED SUBSTANCES

Experiment No.	Test substance	Carboxyl content (m. equiv./100 g)	Copper No.	Mean uptake by slices (% of control value)	
				Strontium-85	Calcium-45
1	Alginate	—	3	65	90
	Polygalacturonate	—	7	86	84
	"Oxycel"	369	52	103	101
	"Surgicel"	354	37	93	93
2	Alginate	—	3	71	90
	Pectin	—	34	88	85
3	Alginate	—	3	68	91
	Amylose	1	0·5	99	108
	Oxidize amylose (*a*)	624	55	87	83
	(*b*)	717	66	77	78
	(*c*)	720	60	97	87
	(*d*)	758	68	90	85
	(*e*)	806	66	82	80

In each experiment the uptake of calcium-45 and strontium-85/g wet tissue is expressed relative to controls incubated without test substances, and is the mean of between two and six flasks of slices.

(From G. Patrick, *Nature, Lond.* **216,** 815 (1967).)

TABLE 2.8 (*cont.*)

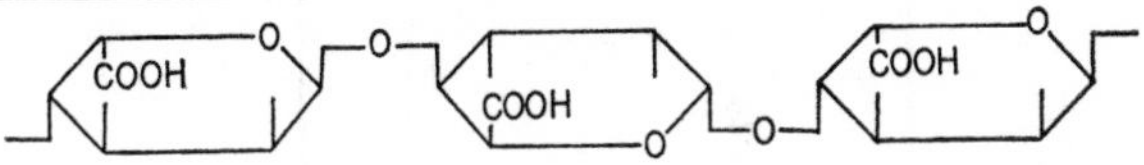

α-1,4-poly-L-guluronic acid

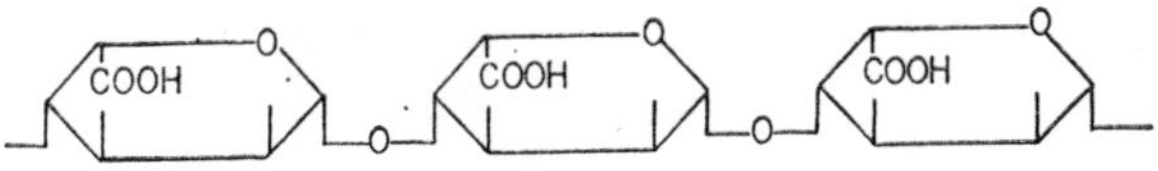

β-1,4-poly-L-guluronic acid

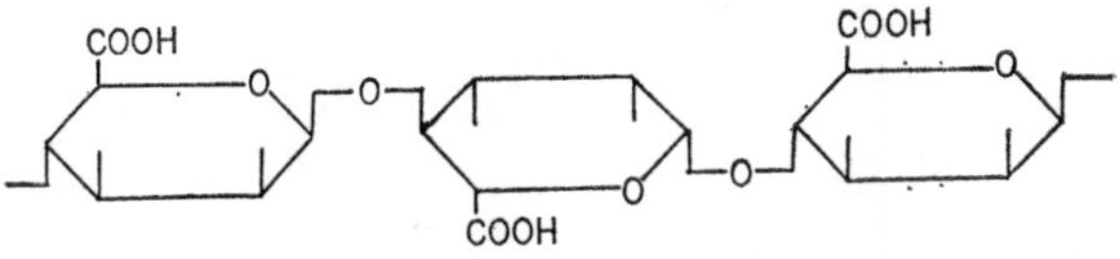

β-1,4-poly-D-mannuronic acid

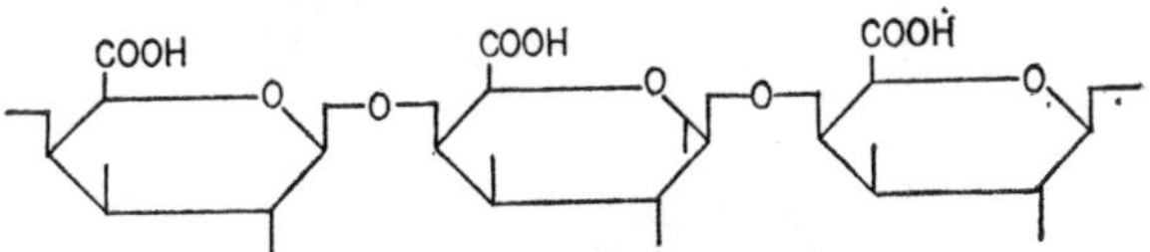

β-1,4-poly-D-galacturonic acid

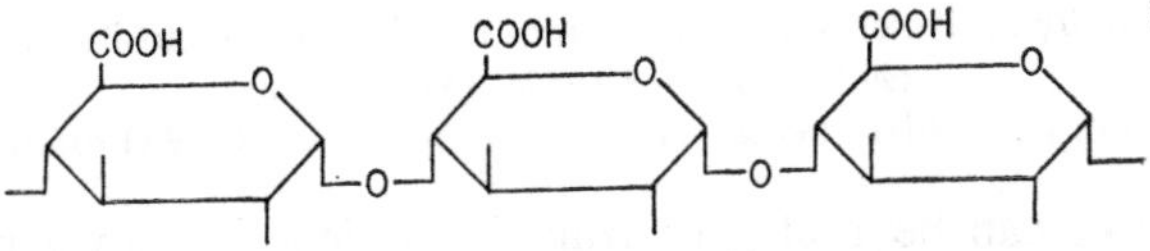

α-1,4-poly-D-glucuronic acid

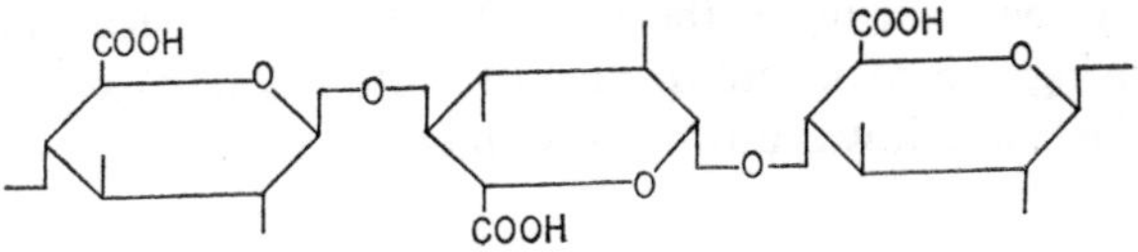

β-1,4-poly-D-glucuronic acid

An incidental human contamination by $^{226}RaSO_4$ was studied by following the faecal excretion after treatment with alginate.[61] It was concluded that sequestration of Ra still in the intestine when the first dose of alginate was given was possible; the continuous presence of alginate in the

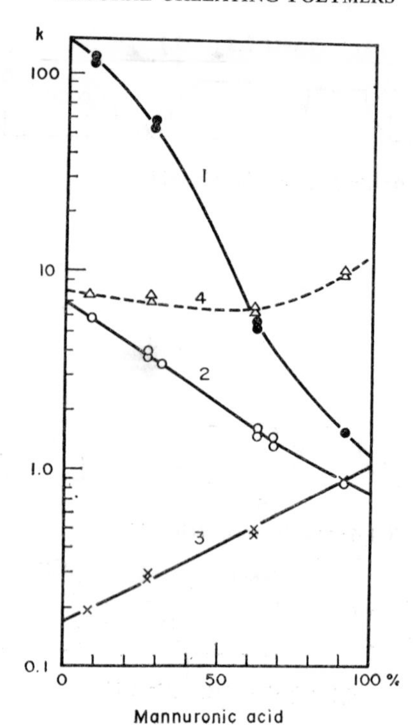

FIG. 2.7. Selectivity coefficients as a function of uronic acid composition: (1) Sr–Mg; (2) Sr–Ca; (3) Co–Ca; (4) Cu–Ca.
(From O. Smidsrod *et al., Acta Chem. Scand.* **22,** 1989 (1968.))

intestinal tract can be useful to trap the radioactive powder inhaled, as long as some Ra can be supposed to be delivered to the gastro-intestinal tract by the clearing of the lung. Most of the ^{226}Ra that entered the body during the exposure to dust of $^{226}RaSO_4$ was excreted during the first 11 days, and it was estimated as high as 90%.

CHROMATOGRAPHY OF METAL IONS ON ALGINATE

Chromatographic techniques involving separation of metal ions on alginate columns are very poorly developed, at the present time. No systematic research has been reported on this topic, and only two simple separations have been described.[62]

Ferric chloride, nitrate, perchlorate and sulphate solutions, acidified with the corresponding acid were tested on 1·2×24 cm columns filled with sodium alginate SS/DJ powder, whose granule diameter was mostly between 0·06 and 0·10 mm, and whose titration with NaOH gave 5·25 meq g^{-1}. After introducing a 25 ml sample, the acid solution, at the desired pH and made 3% with H_2O_2, was used for washing. Preliminary results showed that 0·05 M $HClO_4$ (pH between 1·3 and 1·8) with 3% H_2O_2 at the flow-rate of one ml min^{-1} prevented ferric ions from passing through the column. However the trend for the four acids mentioned, in the pH range 0·6–1·5 is consistent, as at pH = 0·7 98% of iron passes with the first 80 ml, while at higher pH values some iron is also found in further fractions of sulphuric, nitric and hydrochloric acids. In perchloric acid solution at pH = 1·0 no iron is found in the 200 ml effluent, and therefore perchloric acid with 3% H_2O_2 is proposed for the separation of iron from peroxovanadate, which is not retained on the alginate column under these conditions, as reported in Table 2.9.

TABLE 2.9. SEPARATION OF IRON(III) FROM PEROXOVANADATE BY CHROMATOGRAPHY ON A COLUMN OF ALGINIC ACID

Introduced		Found	Error %	Weight ratio
Fe (mg)	V (μg)	V (μg)		Fe/V
10	40	39·6	−1	250
10	30	29·6	−1·3	333
10	23	22·5	−2·2	430
10	20	19·8	−1	500
10	15	14·8	−1·3	666
10	12	11·7	−2·5	833
30	30	29·0	−3·3	1000
20	20	19·8	−1	1000
10	10	9·9	−1	1000
30	10	9·8	−2	3000
40	10	9·6	−4	4000

(From G. Raspi *et al.*, *Ann. Chim.* **58,** 922 (1968).)

One can note, however, that the reported errors on spectrophotometric determinations of vanadium are always negative. Apparently vanadium is not collected to a significant extent because of its chemical form, but no information is available for other chemical forms of vanadium.

Similar treatment has been applied to zinc and indium solutions: the samples contained 5×10^{-5} mol $InCl_3$ or 5×10^{-4} mol $ZnCl_2$ in 25 ml of solution of the acid of interest. The results for indium are strictly similar to those for ferric iron, while for zinc one should reach pH = 1·9 in order to avoid ion leakage and the authors attribute these behaviours to the ionic charges. The distribution coefficients for these two ions are as follows:

Perchloric acid normality	Distribution coefficients K_d	
	Indium	Zinc
0·1	191	0
0·05	350	0
0·01	1150	18
0·001	3020	206

On these grounds the final conditions for the separation of indium from zinc was established as follows: sample volume 50–100 ml, max 0·5 N in $Zn(ClO_4)_2$; eluent 0·1 N perchloric acid, columns $2\cdot5\times17$ cm conditioned with 0·1 N perchloric acid. Flow-rate 4 ml min^{-1}. Zinc passes first in 250 ml. It can be conveniently eluted afterwards with 300 ml of 2 N hydrochloric acid. Polarographic determinations gave the results presented in Table 2.10.

While the reported study is unfortunately strictly limited to the two simple operations described, it emphasizes the possibility of operating a chromatographic column at very low pH and in various acids.

The sodium alginate SS/DJ which is the same alginate used in most of the experiments on strontium and radium body burden reduction and for the column experiments reported above, is a guluronic rich alginate, and has been used as a support for thin layer chromatography, after washing with acids. A powder passed through a 150 mesh sieve was used in water (3·5 g in 20 ml) or in water–methanol 3 : 1 by volume, to cover a 20×20 cm plate with a 200 μ thick layer. 500-μ or thicker layers have poor mechanical properties which also depend on the grain size. However, thickness was found to have little effect on the R_f values. Temperature up to 40 °C

TABLE 2.10. SEPARATION OF INDIUM FROM ZINC BY CHROMATOGRAPHY ON A COLUMN OF ALGINIC ACID

Introduced		Found		Error %		Weight ratio
Zn (mg)	In (mg)	Zn (mg)	In (mg)	Zn	In	Zn/In
228·8	5·70	228	5·80	−0·3	+1·8	40
98·9	2·30	98	2·36	−0·9	+2·6	43
230·0	4·60	229	4·50	−0·4	−2·2	50
165·5	2·90	163	2·93	−1·5	+1·0	57
261·0	2·90	262	2·87	+0·4	−1·0	90
327·7	2·90	322	2·94	−1·7	+1·4	113
326·9	1·15	326	1·18	−0·3	+2·6	284
653·8	2·30	655	2·28	+0·2	−0·9	284
653·8	1·15	655	1·19	+0·2	+3,5	568
1307·6	1·15	1300	1·14	−0·6	−0·8	1136
1961·4	1·15	1950	1·04	−0·6	−9·6	1704
3269·0	1·15	3260	1·00	−0·3	−13·0	2942

(From G. Raspi *et al*., *Ann. Chim.* **58**, 922 (1968).)

affects R_f by 4×10^{-3} cm °C^{-1}, but higher temperatures lead to anomalous values owing to depolymerization of alginic acid.[63] When developing with acid solutions, the movement of the solvent front does not depend on the acid concentration.

To check whether the ionic charge corresponds to the formal charge, a determination based on the method published by Lederer[64] was carried out.[65] By applying the law of mass action to the exchange reaction:

$$xH_R + M^{x+} = M_R + xH^+$$

where M is a metal ion carrying x positive charges and R indicates the polymer, when ion-exchange is the only mechanism of interaction of the ion with the support in the form of paper or thin layer, the slope of the curve:

$$xpH = R_M + \text{constant}$$

$$\text{where } R_M \text{ is } \log\left(\frac{1}{R_f} - 1\right),$$

should be nearly 1·1 for monovalent cations, between 1·4 and 2·1 for divalent cations and between 2·4 and 2·7 for trivalent cations, while it is 3·7 for tetravalent cations. It is therefore possible to determine the number of carboxyl groups involved in the exchange reaction through the R_M values relevant to eluents of different mineral acid concentrations.

It was found that in the interval 0·05–0·5 M acid concentration the plots of R_M vs. the log of the acid concentrations are straight. The slopes are reported in Table 2.11.

One can remark that for Ba and Sr the slopes are well below the expected limit, and this would indicate a mode of interaction where the carboxyl group plays a limited role. This is particularly evident when one considers the reported values for Ca: this again is a point of difference

TABLE 2.11. SLOPES OF THE R_M-LOG [HNO_3] RELATIONSHIP FOR IONS OF VARIOUS IONIC CHARGES ON ALGINIC ACID (AA) AND ACETYLATED ALGINIC ACID (AAA) THIN LAYERS

Ion	Charge	Slope		Ion	Charge	Slope	
		AA	AAA			AA	AAA
Ag	1	0·9	1·0	Co	2	1·5	1·5
Tl	1	0·7	1·0	Zn	2	1·5	1·5
Pb	2	1·3	1·6	Ca	2	1·3	1·5
Ba	2	0·9	1·6	Mg	2	1·4	1·4
Sr	2	0·9	1·5	Ga	3	2·4	2·2
Cu	2	1·4	1·6	In	3	2·4	n. d.
Cd	2	1·4	1·5	Th	4	3·0	n. d.
Ni	2	1·5	1·5				

(From D. Cozzi *et al.*, *J. Chromatog.* **40**, 130–137 (1969).)

between calcium and strontium. For the ions reported in Table 2.10, except Ba and Sr, the interaction involves a number of carboxyl groups equal to the valence of the ion. The values are generally a little lower than expected due to the formation of a pH gradient along the thin layers of both alginic and acetylated alginic acids. Thin layer chromatography data are reported in the Tables 2.12–2.16.[66]

These R_f values deserve some comments: in the II A Group, R_f decreases with increasing atomic number, in agreement with data by other authors cited previously, on the metabolism of these elements. For zinc and indium the trend observed in column chromatography is repeated in thin layer chromatography: this applies also to iron(III) whose R_f is generally low at the acid concentrations studied by other authors in

TABLE 2.12. R_f VALUES OF ELEMENTS ON ALGINIC ACID THIN LAYERS

Eluent	Moles l^{-1}	Cu(II)	Ag(I)	Au(III)	Be(II)	Mg(II)	Ca(II)	Sr(II)	Ba(II)
$(COOH)_2$	0·01	0·00	0·00	0·10±0·10	0·06±0·06	0·06±0·06	n.d.*	0·00	0·00
	0·1	0·00	0·00	0·12±0·12	0·80±0·07	0·72±0·06	n.d.	0·21±0·04	0·08±0·04
	0·5	0·00	0·00	0·14±0·14	0·94±0·06	0·94±0·06	n.d.	n.d.	0·20±0·04
CH_3COOH	1	0·00	0·00	(−)*	0·04±0·04	0·06±0·06	0·00	0·00	0·00
HCl	0·01	0·02±0·02	0·00	0·26±0·09	0·05±0·05	0·09±0·09	0·00	0·00	0·00
	0·1	0·34±0·04	0·00	0·35±0·08	0·63±0·07	0·68±0·06	0·44±0·05	0·32±0·04	0·08±0·04
	1	0·95±0·05	0·00	0·40±0·08	0·93±0·07	0·94±0·06	0·95±0·05	0·68±0·05	0·30±0·04
H_3PO_4	0·01	0·00	0·00	0·14±0·10	0·05±0·05	0·05±0·05	0·00	0·00	0·00
	0·1	0·10±0·06	0·20±0·04	0·16±0·11	n.d.	0·29±0·06	0·12±0·05	0·10±0·05	0·03±0·03
	1	0·70±0·06	0·43±0·04	0·18±0·11	n.d.	0·94±0·06	n.d.	0·48±0·05	0·15±0·04
HNO_3	0·01	0·00	0·07±0·04	0·14±0·10	0·04±0·04	0·08±0·08	0·00	0·00	0·00
	0·1	0·33±0·05	0·42±0·04	0·18±0·09	0·63±0·07	0·67±0·06	0·44±0·05	0·32±0·05	0·08±0·04
	0·5	0·94±0·06	0·64±0·04	0·23±0·10	0·94±0·06	0·94±0·06	0·93±0·07	0·64±0·05	0·25±0·04
$HClO_4$	0·01	0·00	0·06±0·04	0·15±0·10	0·05±0·05	0·08±0·08	0·00	0·00	0·00
	0·1	0·33±0·04	0·37±0·04	0·20±0·12	0·62±0·07	0·68±0·06	0·43±0·05	0·31±0·05	0·07±0·04
	1	0·94±0·06	0·60±0·04	0·30±0·13	0·93±0·07	0·94±0·06	0·95±0·05	0·68±0·05	0·30±0·05
Amount (μg)		5	5	3	0·7	1	7	15	10

* (−) = diffuse. * n.d. = not determined.

(From D. Cozzi *et al.*, *J. Chromatogr.* **35**, 405 (1968).)

TABLE 2.13. R_f VALUES OF ELEMENTS ON ALGINIC ACID THIN LAYERS

Eluent	Moles l⁻¹	Zn(II)	Cd(II)	Hg(II)	Al(III)	Ga(III)	In(III)	Tl(I)	Ge(IV)	Sn(IV)	Pb(II)
$(COOH)_2$	0·01	0·00	0·00	0·66±0·08	0·93±0·07	0·93±0·07	0·15±0·05	0·05±0·04	(–)*	0·00	0·00
	0·1	0·32±0·05	0·33±0·04	0·69±0·07	0·94±0·06	0·94±0·06	n.d.	0·17±0·04	0·94±0·06	0·00	0·00
	0·5	0·96±0·04	0·95±0·05	0·75±0·07	0·94±0·06	0·94±0·06	n.d.	0·25±0·04	0·94±0·06	0·00	0·00
CH_3COOH	1	0·00	0·00	0·64±0·06	0·00	0·00	0·00	0·00	(–)	0·00	0·00
HCl	0·01	0·02±0·02	0·03±0·03	0·79±0·06	0·05±0·05	0·00	0·03±0·03	0·08±0·04	0·94±0·06	0·00	0·00
	0·1	0·58±0·05	0·72±0·05	0·89±0·08	(–)*	0·22±0·06	0·40±0·06	(–)	0·94±0·06	0·00	0·03±0·03
	1	0·94±0·06	0·94±0·06	0·93±0·07	0·94±0·06	0·94±0·06	0·94±0·06	0·00	0·94±0·06	0·00	0·37±0·03
H_3PO_4	0·01	0·00	0·00	0·65±0·05	0·04±0·04	0·00	0·00	0·06±0·04	(–)	0·00	0·00
	0·1	0·14±0·04	0·14±0·04	0·68±0·06	0·24±0·08	0·08±0·06	0·18±0·06	0·09±0·04	(–)	0·00	0·00
	1	0·94±0·06	0·95±0·05	0·72±0·08	0·94±0·06	0·94±0·06	0·94±0·06	0·22±0·04	(–)	0·00	0·06±0·03
HNO_3	0·01	0·02±0·02	0·02±0·02	0·64±0·06	0·05±0·05	0·00	0·00	0·05±0·04	0·94±0·06	0·00	0·00
	0·1	0·54±0·05	0·53±0·04	0·74±0·06	(–)	0·22±0·06	0·03±0·03	0·21±0·04	0·94±0·06	0·00	0·00
	0·5	0·94±0·06	0·95±0·05	0·76±0·06	0·94±0·06	0·93±0·07	0·56±0·07	0·32±0·04	0·94±0·06	0·00	0·10±0·03
$HClO_4$	0·01	0·02±0·02	0·02±0·02	0·63±0·06	0·05±0·05	0·00	0·00	0·05±0·04	0·94±0·06	0·00	0·00
	0·1	0·55±0·05	0·53±0·04	0·73±0·07	(–)	0·21±0·06	0·04±0·04	0·20±0·04	0·94±0·06	0·00	0·00
	1	0·94±0·06	0·95±0·05	0·90±0·10	0·94±0·06	0·94±0·06	0·94±0·06	0·36±0·04	0·94±0·06	0·00	0·14±0·03
Amount (μg)		2·3	1·8	5	0·5	1·5	2·5	10	10	10	5

* (–) = diffuse.

n.d. = not determined.

(From D. Cozzi *et al.*, *J. Chromatogr.* **35**, 405 (1968).)

TABLE 2.14. R_f VALUES OF ELEMENTS ON ALGINIC ACID THIN LAYERS

Eluent	Moles l⁻¹	Ti(IV)	Zr(IV)	Th(IV)	As(V)	As(III)	Sb(III)	Bi(III)	V(V)
$(COOH)_2$	0·01	(−)*	0·04±0·04	0·00	0·94±0·06	0·50±0·04	0·06±0·04	0·06±0·03	0·93±0·07
	0·1	(−)	0·91±0·09	0·00	0·94±0·06	0·49±0·04	0·50±0·04	0·18±0·04	0·94±0·06
	0·5	(−)	0·94±0·06	0·00	0·94±0·06	0·50±0·04	0·73±0·04	0·24±0·06	0·95±0·05
CH_3COOH	1	0·91±0·09	0·00	0·00	0·94±0·06	0·48±0·04	0·00	0·00	0·00
HCl	0.01	0·91±0·09	0·00	0·00	0·94±0·06	0·54±0·04	0·00	0·00	0·03±0·03
	0·1	0·92±0·08	0·05±0·05	0·00	0·94±0·06	0·55±0·04	0·03±0·03	0·03±0·03	0·33±0·05
	1	0·92±0·08	0·06±0·06	0·74±0·08	0·94±0·06	0·52±0·04	0·22±0·03	0·93±0·07	0·95±0·05
H_3PO_4	0·01	(−)	0·00	0·00	0·94±0·06	0·45±0·04	0·00	0·00	0·03±0·03
	0·1	(−)	0·00	0·00	0·94±0·06	0·48±0·04	0·05±0·05	0·00	0·22±0·06
	1	(−)	0·00	0·00	0·94±0·06	0·50±0·04	0·30±0·05	0·00	0·93±0·07
HNO_3	0·01	0·91±0·09	0·00	0·00	0·94±0·06	0·47±0·04	0·00	0·00	0·04±0·04
	0·1	0·91±0·09	0·00	0·00	0·94±0·06	0·48±0·04	0·03±0·03	0·03±0·03	0·30±0·06
	0·5	0·92±0·08	0·03±0·03	0·26±0 07	0·94±0·06	0·48±0·04	0·05±0·05	0·05±0·05	0·95±0·05
$HClO_4$	0·01	0·91±0·09	0·00	0·00	0·94±0·06	0·47±0·04	0·00	0·00	0·04±0·04
	0·1	0·91±0·09	0·00	0·00	0·94±0·06	0·47±0·04	0·00	0·00	0·31±0·06
	1	0·92±0·08	0·05±0·05	0·48±0·10	0·94±0·06	0·47±0·04	0·03±0·03	0·03±0·03	0·95±0·05
Amount (μg)		6	8	15	7	15	4	3	4

* (−) = diffuse.

(From D. Cozzi *et al.*, *J. Chromatogr*. **35**, 405 (1968).)

TABLE 2.15. R_f VALUES OF ELEMENTS ON ALGINIC ACID THIN LAYERS

Eluent	Moles l^{-1}	Cr(III)	Mo(VI)	W(VI)	U(VI)	Mn(II)	Re(III)	Se(IV)	Te(IV)
$(COOH)_2$	0·01	0·06±0·06	0·93±0·07	0·06±0·06	0·56±0·11	0·06±0·06	0·68–0·83	0·95±0·05	0·00
	0·1	0·33±0·05	0·94±0·06	0·08±0·08	0·94±0·06	0·51±0·08	0·68–0·83	0·95±0·05	0·00
	0·5	0·94±0·06	0·95±0·05	0·11±0·07	0·94±0·06	0·93±0·07	0·68–0·83	0·95±0·05	0·05±0·04
CH_3COOH	1	0·00	0·00	0·00	0·00	0·03±0·03	0·68–0·83	0·95±0·05	0·00
HCl	0·01	0·00	0·00	0·06±0·06	0·03±0·03	0·05±0·05	0·68–0·83	0·95±0·05	0·00
	0·1	0·39±0·05	0·06±0·04	0·08±0·08	0·28±0·05	0·64±0·05	0·68–0·83	0·95±0·05	0·05±0·04
	1	0·93±0·07	0·37±0·04	0·08±0·08	0·95±0·05	0·93±0·07	0·68–0·83	0·95±0·05	0·28±0·04
H_3PO_4	0·01	0·04±0·04	(–)	(–)	0·03±0·03	0·05±0·05	0·68–0·83	0·95±0·05	0·00
	0·1	0·08±0·08	(–)	(–)	0·45±0·04	0·10±0·10	0·68–0·83	0·95±0·05	0·00
	1	(–)*	0·93±0·07	0·08±0·06	0·94±0·06	n.d.*	0·68–0·83	0·95±0·05	0·05±0·04
HNO_3	0·01	0·00	0·00	0·05±0·05	0·04±0·04	0·04±0·04	0·68–0·83	0·95±0·05	0·00
	0·1	0·38±0·05	0·07±0·07	0·07±0·07	0·21±0·05	0·63±0·05	0·68–0·83	0·95±0·05	0·05±0·04
	0·5	0·92±0·08	0·22±0·04	0·08±0·08	0·92±0·08	0·93±0·07	0·68–0·83	0·95±0·05	0·16±0·04
$HClO_4$	0·01	0·00	0·00	0·06±0·06	0·03±0·03	0·04±0·04	0·68–0·83	0·95±0·05	0·00
	0·1	0·38±0·05	0·07±0·04	0·08±0·08	0·20±0·06	0·64±0·05	0·68–0·83	0·95±0·05	0·04±0·04
	1	0·94±0·06	0·35±0·05	0·08±0·08	0·93±0·07	0·95±0·05	0·68–0·83	0·95±0·05	0·26±0·04
Amount (μg)		3	5	9	11	2	10	2	3

* (–) = diffuse. * n.d. = not determined.
(From D. Cozzi *et al.*, *J. Chromatogr.* **35**, 405–415 (1968).)

TABLE 2.16. R_f VALUES OF ELEMENTS ON ALGINIC ACID THIN LAYERS

Eluent	Moles l^{-1}	Fe(III)	Co(II)	Ni(II)	Pt(IV)	Ir(IV)	Os(IV)	Rh(III)	Pd(II)
$(COOH)_2$	0·01	(–)*	0·06±0·04	0·13±0·04	0·95±0·05	0·95±0·05	0·95±0·05	(0·00; 0·23; 0·95)	(–)*
	0·1	0·92±0·08	0·93±0·07	0·93±0·07	0·95±0·05	0·95±0·05	0·95±0·05	(0·00; 0·94)	(–)
	0·5	0·93±0·07	0·95±0·05	0·95±0·05	0·95±0·05	0·95±0·05	0·95±0·05	(0·00; 0·94)	(–)
CH_3COOH	1	0·04±0·04	0·00	0·00	0·95±0·05	0·95±0·05	0·95±0·05	(0·00; 0·20; 0·95)	0·06±0·06
HCl	0·01	0·05±0·05	0·05±0·04	0·05±0·04	0·95±0·05	0·95±0·05	0·95±0·05	(0·00; 0·26; 0·95)	0·03±0·03
	0·1	0·12±0·09	0·60±0·05	0·59±0·05	0·95±0·05	0·95±0·05	0·95±0·05	(0·00; 0·95)	(–)
	1	0·93±0·07	0·94±0·06	0·95±0·05	0·95±0·05	0·95±0·05	0·95±0·05	(0·00; 0·95)	(0·00; 0·95
H_3PO_4	0·01	0·04±0·04	0·06±0·04	0·06±0·04	0·95±0·05	0·95±0·05	0·95±0·05	(0·00; 0·28; 0·94)	0·00
	0·1	0·24±0·09	0·13±0·04	0·13±0·04	0·95±0·05	0·95±0·05	0·95±0·05	(0·00; 0·46; 0·95)	0·03±0·03
	1	0·93±0·07	0·95±0·05	0·95±0·05	0·95±0·05	0·95±0·05	0·95±0·05	(0·00; 0·95)	0·08±0·08
HNO_3	0·01	0·05±0·05	0·03±0·03	0·03±0·03	0·95±0·05	0·95±0·05	0·95±0·05	(0·00; 0·26; 0·95)	0·03±0·03
	0·1	0·09±0·09	0·55±0·04	0·55±0·04	0·95±0·05	0·95±0·05	0·95±0·05	(0·00; 0·95)	0·05±0·05
	0·5	(–)	0·95±0·05	0·95±0·05	0·95±0·05	0·95±0·05	0·95±0·05	(0·00; 0·95)	0·12±0·12
$HClO_4$	0·01	0·05±0·05	0·03±0·03	0·03±0·03	0·95±0·05	0·95±0·05	0·95±0·05	(0·00; 0·25; 0·95)	0·03±0·03
	0·1	0·10±0·10	0·54±0·04	0·54±0·04	0·95±0·05	0·95±0·05	0·95±0·05	(0·00; 0·94)	0·05±0·05
	1	0·92±0·08	0·95±0·05	0·95±0·05	0·95±0·05	0·95±0·05	0·95±0·05	(0·00; 0·95)	(–)
Amount (μg)		1·5	1·5	1·5	5	7	5	0·7	1

* (–) = diffuse.

(From D. Cozzi *et al.*, *J. Chromatogr.* **35**, 405 (1968).)

column chromatography. Thallium has low R_f values in spite of its scarce tendency to form complexes, but this is due to the insolubility of the thallium monochloride used. Tin(IV) in no case moved from the original spot because of formation of basic salts, and also lead and zirconium moved scarcely. For the latter it is interesting a comparison with titanium(IV) whose R_f are generally higher than 0·90: this trend was already observed with substituted celluloses, chitin, and chitosan, and it seems that zirconium interacts with these polymers in quite a peculiar way.

The different values for arsenic(III) and arsenic(V) correspond with a double peak elution curve of arsenic from a cellulose column in organic solvents mixed with nitric acid. There is a difference between chromium(III), molybdenum(VI) and tungsten(VI) but it is not so great as for amino-substituted celluloses and chitin and chitosan in acidic media because with these polymers salification and cross-linking take place, and this is impossible with an acidic polymer like alginic acid.

Thorium(IV) and uranium(VI) R_f differ largely in all cases mostly because of uranium complexation.

Selenium(IV) and tellurium(IV) are easily separated by the various eluents. The chromatographic behaviour is in agreement with the non-metallic character of selenium, and with the metallic properties of tellurium.

Manganese is slightly retained in agreement with data on precipitation of alginates.

Platinum(VI), iridium(IV) and osmium(IV) applied to the thin layers as chlorocomplexes, are not retained by alginic acid thin layers. Multiple spots are obtained with rhodium(III).

As far as acetic acid solutions are concerned, only a few ions move from the application point. They are mercury(II), titanium(IV), arsenic (III), arsenic(V), selenium(IV), renium(III), platinum(IV), iridium(IV) and osmium(IV).

A general trend is for an increase of the R_f values with increasing acid concentration.

Alkali metal salts solutions were also used as eluents. In these cases it was noted that ion exchange occurred between bonded hydrogen ions and sodium ions together with the transfer of hydrogen ions towards the solvent front. The formation of a pH gradient has a marked effect on the chromatography as can be seen from Table 2.17. Such a gradient plays the part of increasing the law of mass action and of decreasing the reten-

TABLE 2.17. R_f VALUES OF METAL IONS ON ALGINIC ACID THIN LAYERS

Ion	NaCl (mole/l)			$NaClO_4$ (mole/l)	
	0·05	0·1	0·3	0·1	0·3
Tl(I)	0·13	0·15	(0–0·95)	0·16	0·31
Ag(I)	0·00	0·00	0·00	0·28	0·53
Co(II)	0·21	0·54	0·92	0·53	0·90
Ni(II)	0·20	0·54	0·92	0·52	0·90
Hg(II)	0·71	0·85	0·88	0·71	0·74
Mn(II)	0·33	0·60	0·95	0·59	0·95
Mg(II)	0·32	0·59	0·94	0·60	0·95
Cd(II)	0·27	0·61	0·94	0·39	0·83
Zn(II)	0·20	0·54	0·93	0·51	0·88
Ca(II)	0·06	0·21	n.d.*	0·22	n.d.
Sr(II)	0·04	0·05	0·07	0·05	0·07
Ba(II)	0·00	0·00	0·05	0·00	0·05
Cu(II)	0·02	0·08	0·25	0·07	0·20
Pb(II)	0·00	0·00	0·00	0·00	0·00
Bi(III)	0·00	0·00	(0–0·90)	0·00	0·00
In(III)	0·02	0·02	0·80	0·00	0·00
Th(IV)	0·00	0·00	0·00	0·00	0·00
Mo(VI)	0·00	0·00	0·00	0·00	0·00

* n.d. = not determined.

(From D. Cozzi *et al.*, *J. Chromatogr.* **42**, 532 (1969).)

tion capacity along the layer.[67] Among the Group II A ions which show different affinities for alginic acid, barium and strontium do not respond to the increased salt concentration of the eluent. On the other hand, calcium and magnesium which show the lowest affinity for alginic acid are markedly influenced.

The chromatographic results obtained with thin layers of alginic acid have been compared to results obtained with carboxymethylcellulose and acetylated alginic acid.[63] Data for a few ions are in Table 2.18.

It is clear that the ions considered here have greater R_f values on carboxymethylcellulose in evident connection with the smaller number of carboxyl groups present in carboxymethyl cellulose: in fact its exchange

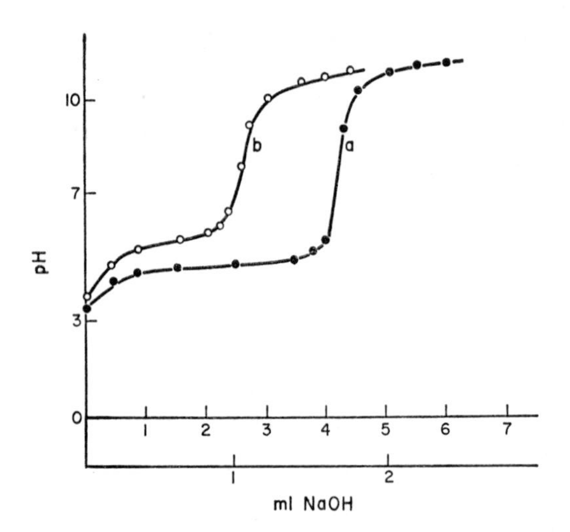

FIG. 2.8. Titration curves of alginic acid (a) and carboxymethylcellulose (b) with 0·1 N sodium hydroxide solution. The expanded scale is referred to carboxymethylcellulose. (From D. Cozzi *et al.*, *J. Chromatogr.* **40**, 130 (1969).)

TABLE 2.18. R_f VALUES ON ALGINIC (AA) AND CARBOXYMETHYLCELLULOSE (CM) THIN LAYERS

Ions	0·01 M HCl		0·1 M HCl		1 M HCl	
	AA	CM	AA	CM	AA	CM
Pb(II)	0·00	0·23±0·04	0·03±0·03	0·92±0·05	0·37±0·03	0·96±0·04
Tl(I)	0·08±0·03	0·70±0·09	(—)	0·80±0·08	0·00	0·00
Bi(III)	0·00	0·00	0·03±0·03	0·50±0·09	0·94±0·06	0·97±0·03
Cu(II)	0·02±0·02	0·46±0·09	0·34±0·04	0·93±0·03	0·95±0·05	0·97±0·03
Fe(III)	0·04±0·04	(—)*	0·12±0·09	0·94±0·03	0·93±0·07	0·97±0·03
Ni(II)	0·05±0·04	0·50±0·04	0·60±0·05	0·96±0·03	0·95±0·05	0·97±0·03
Ba(II)	0·00	0·49±0·06	0·09±0·04	0·92±0·05	0·30±0·04	0·96±0·04

* (−) = diffuse.

(From D. Cozzi *et al.*, *J. Chromatogr.* **35**, 396 (1968).)

capacity is four times smaller than that of alginic acid,[65] as from titration curves in Fig. 2.8.

Data are also available on the comparison of alginic acid and acetylated alginic acid thin layers: on the latter, ions are expected to be unable to form chelates, mainly because hydroxyl groups are no more available and partially because some steric hindrance due to acetyl groups, on the bond between the carboxyl groups and the cation, cannot be excluded. In fact, from R_f values reported in Table 2.19 it can be seen that several ions are

TABLE 2.19. R_f OF SOME ELEMENTS ON ALGINIC ACID (AA) AND ACETYLATED ALGINIC ACID THIN LAYERS (AAA)
Eluent: 0·01 and 0·05 M HNO_3

Ion	AA		AAA (D.A. = 1.7)	
	0·01 M HNO_3	0·05 M HNO_3	0·01 M HNO_3	0·05 M HNO_3
Ag(I)	0·07	0·26	0·39	0·80
Tl(I)	0·05	0·12	0·38	0·80
Cu(II)	0·00	0·16	0·12	0·68
Zn(II)	0·02	0·30	0·24	0·76
Ni(II)	0·03	0·30	0·25	0·77
Pb(II)	0·00	0·00	0·00	0·11
Mg(II)	0·08	0·42	0·30	0·9
Ba(II)	0·00	0·05	0·11	0·67
Ga(III)	0·00	0·06	0·00	0·41
In(III)	0·00	0·00	0·00	0·06

(From D. Cozzi *et al.*, *J. Chromatogr.* **40,** 130 (1969).)

retained on alginic acid much more than on acetylated alginic acid. Cozzi *et al.*[65] state that there is a loss of selectivity as the differences between the R_f values for the various ions on acetylated alginic acid are not so marked as on alginic acid, but this is not evident from the data presented. Hydroxyl groups cannot be considered responsible for selectivity of course, but they are necessary as well as other functional groups for a selective mode of interaction with a metal ion: this has been generally recognized for all types of substituted polysaccharides.

TABLE 2.20. FIRST-ROW TRANSITION AND POST-TRANSITION METAL ION COLLECTION ON 200 mg CARBOXYMETHYLCELLULOSE (CM) AND ALGINIC ACID PER CENT OF THE AMOUNT OF METAL PRESENT IN 50 ml OF 0·44 mM AQUEOUS SOLUTION (ATOMIC ABSORPTION SPECTROMETRY)

	pH	hr	Cr(III)	Cr(VI)	Mn(II)	Fe(II)	Ni(II)	Cu(II)	Zn(II)	As(V)
CM	2·5	1	20	19	0	10	0	76	0	0
		12	11	23	0	0	14	13	0	60
	EDTA	1								
	5·5	1		12	39	10	42	9	41	0
		12		4	29	0	40	67	27	45
	EDTA	1								
Alginic	2·5	1	89	11	36	27	36	31	36	
		12	80	29	34	19	40	75	32	18
	EDTA	1	64	2	7	5	0	0		
	5·5	1	89	13	60	80	77	93	82	
		12		16	58	40	57	88	13	20
	EDTA	1	15	0	18		12			

(R. A. A. Muzzarelli, original results.)

TABLE 2.21. SECOND- AND THIRD-ROW TRANSITION AND POST-TRANSITION METAL ION COLLECTION ON 200 mg CARBOXYMETHYLCELLULOSE AND ALGINIC ACID. PER CENT OF THE AMOUNT PRESENT IN 50 ml OF 0·44 mM AQUEOUS SOLUTION (ATOMIC ABSORPTION SPECTROMETRY)

	pH	hr	Mo(VI	Ag(I)	Sn(II)	Sb tartrate	Hg(II)	Pb(II)
CM	2·5	1		85	100	15	20	10
		12		24	100	10	20	3
	EDTA	1						
	5·5	1		91	100		23	75
		12		55			23	75
	EDTA	1						
Alginic	2·5	1	29	31	24	16	0	95
		12		76	100		4	94
	EDTA	1		30		5		0
	5·5	1	45				0	96
		12		79			10	99
	EDTA	1						0

(R. A. A. Muzzarelli, original results.)

In Tables 2.20 and 2.21 data are reported on the collection of metal ions on carboxymethylcellulose (CM) and alginic acid under the same experimental conditions. These tables are also related to Tables 1.1, 1.2, 5.3,5 .18 and 6.7.

First of all, we can compare the substituted cellulose with alginic acid: for the latter, the collection percentages are generally higher than for CM in accordance with the higher carboxyl group content of alginic acid. Moreover, with alginic acid no regression of the collection percentages with increasing contact time occurs, while this happens with CM, notably in the case of Cu, Ag and Pb at pH = 2·5. Chromate and molybdate are collected to a limited extent as these are acidic polymers, while the basic DE cellulose shows higher capacity for them (Tables 1.1 and 1.2). 0·1 M EDTA prevents the collection of metal ions on alginic acid, the case of Pb being particularly evident, with the exception of Cr(III).

As the data of Tables 2.12–2.16 and Tables 2.20 and 2.21 were obtained with alginic acid from the same supplier, the figures for 0·01–0·1 N HCl solutions can be compared with those at pH = 2·5: it is evident that the very small R_f values of Ag, Pb and Cr(III) (in Tables 2.12, 2.13 and 2.15) correspond to the highest collection percentage for those metals (in Tables 2.20 and 2.21), while the very high R_f value of Hg in Table 2.13 corresponds to nearly no collection of Hg on alginic acid (Table 2.21). The intermediate R_f values for Cu, Zn, Mn and Ni also correlate well with the respective collection percentages.

The trend reported in Table 2.6 for the alginate viscosity maxima also reflects the sequence of the collection percentages: Pb > Cu > Ni > Zn.

Dolmatova and coworkers[68] studied the distribution coefficients of several metal ions on alginic acid by shaking 100 mg of dilute solution for 12 hr at pH = 3 and obtained the following values: Ce(III), 5270; Y(III), 1400; Pb(II), 3580; Cu(II), 1030; Ba(II), 880; Sr(II), 740; Ca(II), 500; Cd(II), 400; Co(II), 370; Ni(II), 300; Mn(II), 260; and Zn(II), 250. Also these data are in agreement with the collection percentages in Tables 2.20 and 2.21.

Therefore the information reached independently by Cozzi, Dolmatova, Haug, Muzzarelli and other authors is reciprocally corroborated and, in addition, confirms that the collection percentages are reliable indications for predicting the chromatographic behaviour of metal ions.

INTERACTIONS WITH METAL IONS, AND SOLUBILITY

Most water-soluble alginates are precipitated by addition of water-miscible organic solvents such as alcohols and ketones. For a complete precipitation, the less polar solvents are required in smaller amounts that depend on the organic or inorganic cations being present in the following order:

easier precipitation
←
Mg, Na, K, ammonium

In numerous applications use is made of the insolubilization of alginates by the addition of a calcium salt. This precipitation reaction can be explained with crosslinkage through the divalent ions involving carboxyl groups of different chains and thus forming a macromolecule: after sufficient growth gelation or precipitation occurs.[69, 70] An alternative simple reason is the mere insolubility of the salt formed, as in the case of the silver salt where the type of crosslinkage mentioned above is out of the question.

Data on partially acetylated alginates have been brought forth to explain further the interactions of metal ions with the polymer. Ammonium diacetyl alginate is compatible with calcium and other divalent ions: it is therefore clear that the carboxyl groups are only partially responsible for the precipitation. The presence of free hydroxyl groups is necessary and apparently gelation occurs by crosslinkage through chelate formation involving both carboxyl and hydroxyl groups.

Not only diacetyl alginates are compatible with metal ions but also partially acetylated alginates. A product with a D.A. of about 1·4 consists exclusively of mono and diacetate units. However, in the presence of free hydroxyl groups belonging to the monoacetylated units, no gelation could be noticed on addition of calcium ions. The chelate structure of non acetylated calcium alginate, proposed by Schweiger, should therefore be as presented in Fig. 2.9 and its first requirement is the availability of two adjacent hydroxyl groups. Molecular models permit one to form links in this way with both mannuronic and guluronic acid units, but the chains would be very distorted. X-ray diffraction spectra of calcium alginate are not very sharp to discriminate among the various possibilities, but a

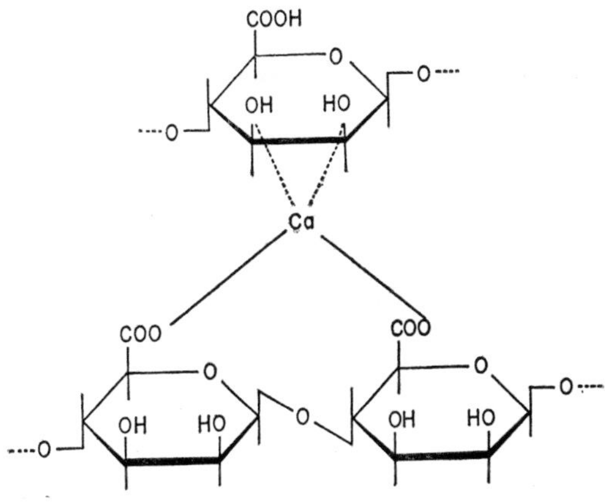

FIG. 2.9.

TABLE 2.22.

D.A.	Viscosity (cp)	Consistency and/or viscosity (cp) after addition of 0·25 g $CaCl_2$
0·05	171	Precipitates*
0·10	134	
0·27	100	
0·48	91	Gelatinous, lumps
0·85	116	Heavy gel
0·98	61	~2000, thin gel
1·09	119	Gel
1·18	108	Soft gel, ~3800
1·21	34	700–900, thin gel
1·39	71	75
1·42	64	43
1·47	55	37
1·51	100	100
1·59	98	54
1·76	43	28

* With increasing D.A. increasing amounts of $CaCl_2$ are necessary for precipitation.

(From R. G. Schweiger, *J. Org. Chem.* **27**, 1789 (1962).)

TABLE 2.23.

D.A.	Viscosity (cp)	Consistency and/or viscosity (cp) after addition of $ZnCl_2$ 0·25 g	$ZnCl_2$ 0·50 g	$ZnCl_2$ 0·75 g	$ZnCl_2$ 1·00 g	$BaCl_2$ 0·25 g	$BaCl_2$ 0·50 g	$BaCl_2$ 0·75 g	1 g $MnCl_2$ or 1 g $MgCl_2$
0·6	46	~1420, gel, lumps	~1150	500, thinning out	310	Heavy gel	Thinning out		31
0·7	39	~1180, gel, lumps	~1130	620	380	Heavy gel	Thinning out		
0·87	47	~600, thin gel	790	540	360	Heavy gel	Heavy gel	Thinning out	30
0·98	61	410	390	390	240	Gel, ~4850	~3300	~2400	
1·21	45	99	133	107	89	Gel, ~3700	~2950	~2400	24
1·39	70	117	134	121	107	Solution 1240	1160	1050	51
1·42	66	48	54	54	52	140–190	160–190	150	
1·47	55	32	31	32	31	31	30	29	

(From R. G. Schweiger, *J. Org. Chem.* **27**, 1789 (1962).)

further explanation for the absence of precipitates of partially acetylated alginates with metal ions could be that, for steric reasons, acetyl groups themselves prevent the polymer chains from coming close together enough for crosslinking. In any case the experimental evidence is in Table 2.22.

Similar measurements were made with zinc, barium, manganese, and magnesium,[70] and results are in Table 2.23. The same observations can be made. The critical D.A. value is again very close to the value for calcium, a little higher than 1. Manganese and magnesium do not form complexes and thus do not react with partially acetylated alginate. Zinc and barium are very effective in forming gels, barium being more effective than calcium. No quantitative data are reported for strontium but even so the trend agrees with data obtained by other techniques with acetylated and non acetylated alginic acid.

On the basis of the chromatographic experiments it has been observed that when a metal salt solution is added to a suspension of alginic acid, the pH decreases and the magnitude of the decrease depends upon the affinity of the metal ion for alginic acid.[65] The increase of acidity was therefore used for the following purposes:

(i) to evaluate the affinity of different ions with respect to the acetylated and non-acetylated alginic acids;
(ii) to see whether it is possible to correlate the data obtained in such a way with the chromatographic data;
(iii) to understand the ion-exchange mechanism by the comparison under the same experimental conditions, of the data obtained on alginic acid with that on the acetylated alginic acid.

Table 2.24 gives the pH values obtained after the addition of metal salt solutions to aqueous suspensions of samples of alginic acid, acetylated alginic acid with D.A. 0·93 and 1·7, and carboxymethylcellulose. From the above data an affinity scale for the ions with respect to the polymer was written as follows:[65]

Alginic acid

(a) Ag > Tl, Cs > K > Na > Li.
(b) Pb > Ba > Sr > Cu, Cd > Ca > Be, Zn, Co, Ni > Mn, Mg.
(c) In > Ce > Cr.

Acetylated alginic acid (D.A. = 0·93)

(a) Ag > Cs, Tl > K, Na, Li.
(b) Pb > Cu, Cd, Ba > Sr > Ca > Be, Zn, Co, Ni, Mn > Mg.
(c) In > Ce > Cr.

Acetylated alginic acid (*D.A.* = 1·7)

(a) Ag > Cs > Tl, Li.
(b) Pb > Cu, Cd > Ba, Sr, Ca, Be, Zn, Co, Ni, Mn > Mg.
(c) In > Ce, Cr.

Carboxymethylcellulose

(a) Ag > Cs > Tl, Li.
(b) Pb > Cu, Cd, Ba > Ca, Sr > Zn, Co, Ni, Mn > Be, Mg.
(c) In > Ce > Cr.

TABLE 2.24. pH OF AQUEOUS SUSPENSIONS OF ALGINIC ACID (AA) (INITIAL pH = 2·97), ACETYLATED ALGINIC ACID (AAA) WITH D.A. 0·93 (INITIAL pH = 3·32) AND 1·7 (INITIAL pH = 3·84) AND CARBOXYMETHYLCELLULOSE (CMC) (INITIAL pH = 3·21) AFTER THE ADDITION OF METAL SALT SOLUTIONS (MEQUIV, SALT/MEQUIV. EXCH. = 1·43)

Metal salt	AA	AAA (0·93)	AAA (1·7)	CMC
$AgNO_3$	2·42	2·67	2·72	2·82
Tl_2SO_4	2·64	2·80	2·86	3·06
CsCl	2·66	2·80	2·82	2·94
KCl	2·72	2·89	—	—
NaCl	2·80	2·90	—	—
LiCl	2·90	2·90	2·91	3·06
$Pb(NO_3)_2$	2·04	2·24	2·53	2·20
$Ba(NO_3)_2$	2·20	2·48	2·77	2·56
$Sr(NO_3)_2$	2·34	2·52	2·78	2·67
$CuSO_4$	2·40	2·47	2·59	2·56
$Cd(NO_3)_2$	2·41	2·46	2·57	2·58
$Ca(NO_3)_2$	2·44	2·68	2·78	2·66
$BeSO_4$	2·60	2·74	2·76	2·88
$ZnSO_4$	2·61	2·76	2·78	2·85
$CoSO_4$	2·61	2·75	2·78	2·84
$NiSO_4$	2·62	2·76	2·78	2·84
$MnSO_4$	2·72	2·78	2·80	2·85
$MgSO_4$	2·74	2·88	2·96	2·88
$InCl_3$	1·98	2·18	2·23	2·28
$Ce(NO_3)_2$	2·12	2·36	2·46	2·54
$CrCl_2$	2·18	2·41	2·48	2·60

D.A. = Degree of acetylation.

(From D. Cozzi *et al.*, *J. Chromatogr.* **40**, 130 (1969).)

For alginic acid the conclusion was reached that for homologous ions the affinity for alginic acid correlates with the size of the hydrated ionic radius and decreases in the following order:

$$Cs^+ > K^+ > Na^+ > Li^+$$
$$Ba^{++} > Sr^{++} > Ca^{++} > Mg^{++}$$
$$Cd^{++} > Zn^{++}$$

These results agree with those for carboxymethylcellulose reported by other authors.[71]

The monovalent ions do not cross link as the viscosity measurements show: the same can be said for thallous sulphate. Unlike the alkali ions, Ag(I) and Tl(I) which form insoluble alginates in some instances. Ag(I) forms a precipitate with sodium alginate and ammonium acetyl alginate with D.A. 0·93 but not with alginate with D.A. 1·7, although in both cases affinity is the same as from Table 2.24. This behaviour would indicate that the presence of intramolecular chelates is still possible with alginic acid having D.A. 0·93.

In Table 2.25 information on the precipitation reactions with metal ions is summarized: it seems that the greater the tendency to coordination, the greater the precipitation yield.

In fact the divalent ions were divided into three groups:[65]

1. Ba, Sr, Ca, Co, Ni, and Zn: their behaviour is in agreement with the formation of intermolecular chelates as indicated by Schweiger:
2. Pb, Cu, and Cd: the type of precipitates and the pH data indicate the presence of crosslinkages between the molecules. Such results are in agreement with the viscosity measurements by Haug[38] but for Pb, disagree with those of Schweiger.[72]
3. Mn, Mg, and Be: for these elements the formation of cross linkages is excluded. With Be, however, it is possible to obtain a precipitation, possibly because of formation of basic salts.

Tri- and tetravalent ions seem to bind respectively three and four carboxyl groups in the pH range where the chromatographic experiments have been carried out. As gelatinous precipitates are obtained even with highly acetylated alginates, it is reasonable to assume that carboxyl groups of different chains participate in the reaction with tri- and tetravalent ons. The formation of alginates of Al, Fe(III), and Sn(IV) may be affected by the presence of hydroxylated species of the ions as underlined by the absence of a viscosity maximum in the suspensions of these alginates.

TABLE 2.25. PRECIPITATION REACTIONS WITH METAL IONS*

Metal ion	D.A.		
	0	0·93	1·7
Ag(I)	+	+	—
Tl(I)	+	—	—
Cs(I)	—	—	—
Pb(II)	++	++	++
Cu(II)	++	++	++
Cd(II)	++	++	++
Ba(II)	++	++	—
Sr(II)	++	++	—
Ca(II)	++	++	—
Zn(II)	++	++	—
Co(II)	++	+	—
Ni(II)	++	+	—
Be(II)	+	+	—
Mn(II)	+	—	—
Mg(II)	—	—	—
In(III)	++	++	++
Ce(III)	++	++	++
Cr(III)	++	++	++
Th(IV)	++	++	++

*++ = gelatinous precipitate;
+ = grainy gel; − = no reaction.

(From D. Cozzi *et al.*, *J. Chromatogr.* **40**, 130 (1969).)

CHROMATOGRAPHY OF ORGANIC SUBSTANCES ON ALGINIC ACID

As alginic acid is a polymer carrying carboxyl groups it should interact by ion-exchange with basic substances like amines, amino acids, pyrimidines, and purines. While the results obtained with these substances do not characterize alginic acid from the standpoint of its chelating ability, some of them are reported below to stress the possibilities of uses of this polymer for separations of many substances other than metal ions.

The publications cited here are quite recent, and this again points out that for too many years the natural polymers have been disregarded, in spite of their evident peculiar characteristics which would have qualified them as very interesting chromatographic supports.

It is possible to foresee, for instance, that natural chelating polymers are suitable supports for ligand exchange chromatography, but at present no data is available on this topic, because the interest in these polymers from the chromatographic point of view is too recent.

Amino acids have been studied on thin layers of alginic acid.[73] A mixture of 6 g of alginic acid with 1·5 g of cellulose suspended in 40 ml of water was used to coat four 20×20 cm plates, with a thickness of 300 μ. Each amino acid was dissolved separately to form a 1% solution in 10% isopropanol; tyrosine and cystine required 0·1 N HCl. The plates were developed to 11 cm in about half an hour, by the ascending technique with the eluents mentioned in Table 2.26. The amino acids were visualized with the reagent of Moffat and Lytle. Some interesting separations were obtained as from Table 2.26: for instance, α-alanine and β-alanine and some isomeric aminobutyric acids were separated, thus demonstrating the resolving power of alginic acid, as does the separation of glutamic from aspartic acid: these separations are generally unsuccessful on other chromatographic supports. The high R_f values of taurine were expected on the ground of its marked acidity.

The use of the expression pH $= R_M +$const. gave a curve for tryptophan on alginic acid which shows that adsorption is not determined by ion exchange alone, another factor may be steric hindrance, due to the side chains of the amino acids and the adjacent —OH groups of alginic acid, which influences the reaction between the ionic groups of the amino acid and the adsorbent.

The R_f values for some representative amino acids in 0·05 N HCl differ according to whether the $—NH_2$ group is attached to a primary, a secondary or a tertiary carbon atom. These differences are attributable to differences in the steric hindrance of the side chains. Should this observation be of general validity, as further experiments might prove, substances with similar acidities or basicities but different degrees of steric hindrance could be separated.[73] The comparison of alginic acid with carboxymethylcellulose for the amino acid separations, is in favour of alginic acid.

The above mentioned expression relating pH and R_f was applied to

TABLE 2.26. R_f VALUES OF AMINO ACIDS ON ALGINIC ACID THIN LAYERS

Amino acids	Eluent							Amount (μg)
	1	2	3	4	5	6	7	
Arg	0·01	0·00	0·09	0·07	0·02	0·10	0·00	2·0
Lys	0·02	0·00	0·12	0·07	0·02	0·14	0·02	2·0
Orn	0·02	0·00	0·12	0·04	0·03	0·14	0·02	2·0
His	0·02	0·00	0·11	0·04	0·03	0·12	0·02	2·7
$(Cys)_2$	0·04	0·00	0·12	0·02	0·06	0·16	0·07	1·7
Gly (NH_2)	0·12	0·09	0·37	0·23	0·12	0·39	0·09	1·5
Try	0·14	0·33	0·32	e.s.	0·14	0·28	0·13	2·5
γ-AnB	0·14	0·13	0·51	0·49	0·14	0·50	0·10	3·0
β-AiB	0·17	0·16	0·58	0·57	0·16	0·54	0·11	3·5
β-Ala	0·17	0·14	0·51	0·37	0·16	0·50	0·10	1·5
Gly (OCH_2CH_3)	0·19	0·18	0·56	0·48	0·18	0·50	0·13	2·0
Cit	0·20	0·16	0·43	0·39	0·22	0·43	0·19	2·0
Dopa	0·22	0·23	0·46	0·62	0·22	0·46	0·22	1·3
Tyr	0·23	0·29	0·54	0·74	0·23	0·54	0·23	0·8
Gly	0·19	0·09	0·49	0·37	0·23	0·49	0·19	1·5
Glu	0·28	0·20	0·58	0·58	0·28	0·58	0·27	0·4
Glu (NH_2)	0·28	0·16	0·58	0·57	0·28	0·54	0·27	0·5
Phe	0·30	0·36	0·58	0·74	0·30	0·58	0·27	1·3
Ser	0·30	0·15	0·59	0·50	0·30	0·59	0·27	1·2
Met	0·30	0·30	0·59	0·68	0·31	0·58	0·28	1·2
Ala	0·33	0·24	0·60	0·53	0·33	0·61	0·26	1·3
α-AnB	0·34	0·26	0·64	0·64	0·34	0·65	0·29	1·0
Thr	0·34	0·21	0·64	0·60	0·35	0·66	0·30	1·7
Sar	0·32	0·22	0·62	0·57	0·35	0·62	0·32	3·0
Asp	0·33	0·22	0·58	0·50	0·35	0·60	0·33	2·0
Val	0·36	0·35	0·68	e.s.	0·36	0·68	0·29	0·9
α-AiB	0·36	0·29	0·74	0·77	0·36	0·72	0·32	3·5
Ile	0·35	0·42	0·69	e.s.	0·36	0·69	0·35	1·2
Pro	0·37	0·31	0·62	0·56	0·40	0·66	0·40	3·0
β-Cl-Ala	0·46	0·22	0·63	0·47	0·47	0·64	0·44	2·3
Tau	0·85	0·54	0·90	0·56	0·96	0·97	0·86	5·0

e.s. = elongated spot.

Eluents: (1) 0·01 M HCl in 10% isopropanol; (2) 0·01 M HCl in 50% Isopropanol; (3) 0·05 M HCl in 10% isopropanol; (4) 0·05 M HCl in 50% isopropanol; (5) 0·01 M HCl; (6) 0·05 M HCl; (7) 1 M acetic acid.

(From D. Cozzi *et al.*, *J. Chromatogr.* **40**, 138 (1969).)

purines and pyrimidines on thin layers of alginic acid.[74] For most bases the slopes of the curves are between 0·9 and 1·0 which is in accordance with an ion-exchange mechanism involving monovalent cations. For some other bases, like xantine, hypoxantine, and 2,4,6-triaminopyrimidine slightly curvilinear trends are observed, and they were attributed to different protonation of these bases in the pH range explored.

In order to understand the influence of the acid–base characteristics on the retention of these compounds an attempt was made to correlate the R_f values reported in Table 2.26 with the pK_a of the bases. The sequences of R_f and pK_a are in agreement only for the bases whose pK_a values are $\leq 3{\cdot}3$ except at the low pH values of the eluent. As the acidity increases, the difference in the R_f values tends to decrease owing to almost complete protonation of the bases; for the same reason there is no differentiation in the whole pH range among the R_f values of the bases whose pK_a is higher than 3·3.[74] Typical results are presented in Table 2.27.

2,4-Lutidine and 2,4,6-collidine show an increase in the R_f values as the pK_a increases: this is justified by the increased steric hindrance of the nitrogen atom in the ring due to the introduction of methyl groups in the α position. By considering the R_f values obtained with solutions at pH = 1·45 and 1·20 the following sequence for the retention of the bases by the exchanger was obtained: purines > pyrimidines > aromatic amines > pyridines. Such a sequence is the same as the one observed on cellulose with water and 1 M acetic acid as eluents; this may be explained by considering that on cellulosic exchangers the compounds with the largest number of polar groups or centres are the most strongly retained. The sequence for the number of polar centres is as follows: purines > pyrimidines > pyridines = aromatic amines. The stronger retention of aromatic amines in comparison with that of pyridine may be ascribed to the greater possibility of the $—NH_3^+$ groups binding to the carboxyl groups of the polymer than the protonated nitrogen of the pyridine ring.

Primary aromatic amines have been studied in detail on thin layers and on columns of alginic acid.

For thin layer chromatography the following systems were used: hydrochloric acid–acetone, hydrochloric acid–dioxan, chloroacetic acid–dioxan and chloroacetic acid–isopropanol: the latter brought about interesting differences in the R_f values of the isomers. In the case of aminobenzoic acid, for instance, it lowered the R_f value of the meta-isomer and raised that of the other two to permit a separation which could not be achieved

TABLE 2.27. R_f VALUES OF PURINES, PYRIMIDINES, NUCLEOSIDES, PYRIDINES, AND PRIMARY AROMATIC AMINES ON THIN LAYERS OF ALGINIC ACID

Substance	Water	pH of acetic acid and hydrochloric acid solutions					
		2·55	2·20	2·00	1·75	1·45	1·20
Purine	0·06	0·12	0·16	0·23	0·31	0·47	0·60
2-Aminopurine	0·00	0·05	0·09	0·14	0·24	0·37	0·49
Guanine	0·00	0·06	0·10	0·15	0·25	0·38	0·50
2-Amino-6-chloropurine	0·00	0·06	0·10	0·15	0·24	0·37	0·50
Adenine	0·00	0·05	0·10	0·15	0·24	0·37	0·51
Hypoxanthine	0·17	0·24	0·28	0·31	0·39	0·48	0·58
Xanthine	0·40*	0·53*	0·53*	0·53*	0·55*	0·58*	0·61*
Pyrimidine	n.d.	n.d.	n.d.	n.d.	n.d.	n.d.	n.d.
2-Aminopyrimidine	0·01	0·07	0·14	0·21	n.d.	n.d.	n.d.
Isocytosine	0·00	0·07	0·14	0·22	0·32	0·48	0·60
2-Amino-4,6-dihydroxy-pyrimidine	0·31	0·44	0·44	0·45	0·47	0·54	0·63
2-Amino-4,6-dimethyl-pyrimidine	0·00	0·06	n.d.	n.d.	n.d.	n.d.	n.d.
Cytosine	0·00	0·07	0·14	0·22	0·32	0·48	0·60
5-Methylcytosine	0·00	0·07	0·14	0·21	0·31	0·48	0·60
Uracil	0·79	0·81	0·81	0·81	0·82	0·82	0·82
Thymine	0·79	0·80	0·81	0·81	0·81	0·82	0·82
4-Aminouracil	0·52	0·58	0·58	0·58	0·60	0·60	0·61
5-Aminouracil	0·00	0·09	0·11	0·17	0·25	0·45	0·60
4,5-diaminopyrimidine	0·00	0·05	0·10	0·16	0·26	0·43	0·58
2,4-diamino-6-chloropyrimidine	0·01	0·07	0·10	0·16	0·26	0·43	0·59
2,4,6-triaminopyrimidine	0·00	0·02	0·04	0·06	0·11	0·22	0·35
Adenosine	0·00	0·09	0·14	0·22	0·32	0·49	0·61
Guanosine	0·13	0·20	0·24	0·30	0·38	0·49	0·59
Cytidine	0·00	0·08	0·16	0·23	0·34	0·50	0·61
Pyridine	0·00	0·12	0·23	0·32	0·44	0·62	n.d.
2-Aminopyridine	0·00	009	0·18	0·26	0·37	0·53	0·68
4-Picoline	0·00	0·12	0·23	0·31	0·44	0·63	0·74
2,4-Lutidine	0·01	0·14	0·26	0·36	0·49	0·66	0·77
2,4,6-Collidine	0·01	0·16	0·31	0·42	0·56	n.d.	n.d.
Nicotinic acid	0·24	0·26	0·32	0·43	0·56	0·71	n.d.
Nicotinamide	0·01	0·09	0·16	0·24	0·33	0·50	0·64
Pyridoxine	0·00	0·12	0·23	0·33	0·44	n.d.	n.d.
Aniline	0·00	0·10	0·18	0·26	0·38	0·56	0·68
p-Toluidine	0·00	0·09	0·18	0·26	0·38	0·55	0·68
p-Nitroaniline	0·34	0·49	0·50	0·50	0·52	0·53	0·55
p-Chloroaniline	0·00	0·08	0·15	0·22	0·33	0·50	0·64
p-Bromoaniline	0·00	0·08	0·15	0·22	0·33	0·56	0·63
p-Aminobenzoic acid	0·08	0·17	0·20	0·28	0·38	0·53	0·66
p-Aminohippuric acid	0·11	0·16	0·18	0·28	0·39	0·55	0·68
a-Naphthylamine	0·00	0·06	0·11	0·17	0·24	0·40	0·53

* Elongated spot.

Water and acetic acid + hydrochloric acid solutions as eluents. Acetic acid concentration, 1 mole/l. n.d. = not determined.

(From L. Lepri *et al.*, *J. Chromatogr.* **64**, 271 (1972).)

TABLE 2.28. R_f VALUES OF SOME AROMATIC AMINES

Amine	ON ALGINIC ACID Eluent 1	2	3	4	Amount (μg)	ON CARBOXY-METHYLCELLULOSE Eluent water	1	Amount (μg)
m-Aminobenzoic-acid	0·10	0·21	0·48	0·62	0·6	0·14	0·44	0·3
o-Aminobenzoic acid	0·23	0·30	0·55	0·67	0·6	0·44	0·57	0·3
p-Aminobenzoic acid	0·16	0·24	0·51	0·61	0·4	0·28	0·45	0·2
Sulphanilic acid	0·95	0·94	0·94	0·96	2·0	0·96	0·96	0·4
o-Arsanilic acid	0·82	0·78	0·86	0·81	2·0	0·92	0·91	0·6
p-Arsanilic acid	0·41	0·42	0·64	0·72	1·5	0·85	0·81	0·5
m-Aminophenol	0·08	0·18	0·49	0·60	0·6	0·01	0·38	0·5
o-Aminophenol	0·09	0·18	0·51	0·60	1·2	0·01	0·40	0·5
p-Aminophenol	0·08	0·18	0·51	0·60	1·0	0·00	0·40	0·5
α-Naphthylamine	0·06	0·12	0·34	0·40	1·2	0·00	0·28	0·6
β-Naphthylamine	0·06	0·12	0·34	0·40	1·2	0·00	0·26	0·6
m-Anisidine	0·08	0·18	0·53	0·62	1·3	0·02	0·38	0·6
o-Anisidine	0·10	0·21	0·56	0·67	1·3	0·02	0·40	0·6
p-Anisidine	0·08	0·18	0·52	0·62	0·6	0·01	0·38	0·3
Benzidine	0·00	0·00	0·04	0·13	0·3	0·00	0·08	0·3
m-Bromoaniline	0·08	0·16	0·44	0·56	10·0	0·03	0·40	2·0
o-Bromoaniline	0·09	0·19	0·51	0·63	15·0	0·03	0·52	3·0
p-Bromoaniline	0·06	0·16	0·44	0·58	1·2	0·02	0·38	0·5
m-Chloroaniline	0·08	0·18	0·45	0·58	10·0	0·04	0·40	2·0
o-Chloroaniline	0·13	0·25	0·48	0·64	15·0	0·04	e.s.*	3·0
p-Chloroaniline	0·08	0·18	0·43	0·60	0·8	0·03	0·38	0·5
m-Phenylenediamine	0·01	0·02	0·08	0·28	1·0	0·01	0·18	0·5
o-Phenylenediamine	0·06	0·09	0·20	0·63	1·2	0·05	0·35	1·0
p-Phenylenediamine	0·00	0·02	0·08	0·27	0·5	0·00	0·16	0·3
m-Nitroaniline	0·11	0·21	0·40	0·59	0·6	0·25	0·49	0·6
o-Nitroaniline	0·55	0·52	0·62	0·51	2·0	0·52	0·52	2·0
p-Nitroaniline	0·47	0·47	0·56	0·52	0·8	0·47	0·55	0·8
m-Toluidine	0·08	0·25	0·50	0·66	8·0	0·02	0·40	1·5
o-Toluidine	0·08	0·26	0·50	0·66	8·0	0·03	0·44	1·5
p-Toluidine	0·07	0·25	0·50	0·66	1·2	0·01	0·41	0·3
p-Aminodimethylaniline	0·00	0·04	0·14	0·36	2·0	0·01	0·24	1·0
p-Aminosalicylic acid	0·26	0·27	0·46	0·53	1·2	0·50	0·42	0·5
p-Aminoacetophenone	0·20	0·29	0·51	0·67	0·6	0·35	0·58	0·5

* e.s. = elongated spot.

Eluents: (1) 1 N acetic acid; (2) 1 N formic acid; (3) 1 N chloroacetic acid; (4) 0·1 N hydrochloric acid.

(From D. Cozzi *et al.*, *J. Chromatogr.* **43**, 463 (1969).)

with the other systems.[75] Data are reported in Table 2.28, for alginic acid and carboxymethyl cellulose: with the same eluent, namely 1 N acetic acid, the R_f values are higher on carboxymethylcellulose, as one can reasonably expect on the basis of the smaller number of carboxyl groups present in the modified cellulose.

The R_f values obtained on both types of plates can be correlated with the acid–base character of the isomers. For the aminobenzoic acids and nitroanilines, the affinity for the chromatographic support decreases as the pK_b value increases. The pK_b values of *m*-, *p*- and *o*-aminobenzoic acids are 11·44, 11·80 and 12·20 respectively, and correspondingly the R_f values for alginic acid and carboxymethylcellulose increase as $m < p < o$. The same can be said for *m*-, *p*- and *o*-nitroanilines whose pK_b values are 11·54, 13·00 and 14·26 respectively. The smaller affinity of the *o*-isomer with respect to the *p*- and *m*-isomers for aminobenzoic acids, nitroanilines, chloroanilines, bromoanilines, phenylenediamines on both polymers and also toluidines on carboxymethylcellulose and anisidines on alginic acid, suggests that steric hindrance affects the reaction between the groups of the amines and the chromatographic support.

In electrophoresis experiments the retentive power of alginic acid is the controlling factor, because migration on alginic acid is three times as slow as on carboxymethylcellulose. For the latter, which is known from the above presented data to have a weaker retentive power than alginic acid, the electrophoretic results are mostly governed by ionic mobility. High-voltage electrophoresis enables one to separate alginic acid amines that differ very little in their acidity or basicity.

Experiments were also carried out with columns of alginic acid and carboxymethylcellulose. Alginic acid was used in form of a 50–150 mesh powder and 4 g of this polymer were used to fill a column with i.d. 1·1 cm. Two grams of carboxymethyl cellulose treated with 1 N HCl and washed with water were used to fill a similar column that had a flow-rate of 1 ml min^{-1}, while the alginic acid column had a flow-rate of 2 ml min^{-1}.

On alginic acid columns, aromatic diamines in 1 N acetic acid show greater affinity for alginic acid than aromatic monoamines in accordance with the thin layer chromatography data: of course, the possibilities of separation are enhanced in the column, and, for example, *o*-phenylenediamine can be clearly separated from naphthylamines.

From the elution curves presented in Fig. 2.10, the authors[76] deduced that the resolving power of a column is less with respect to that of a thin

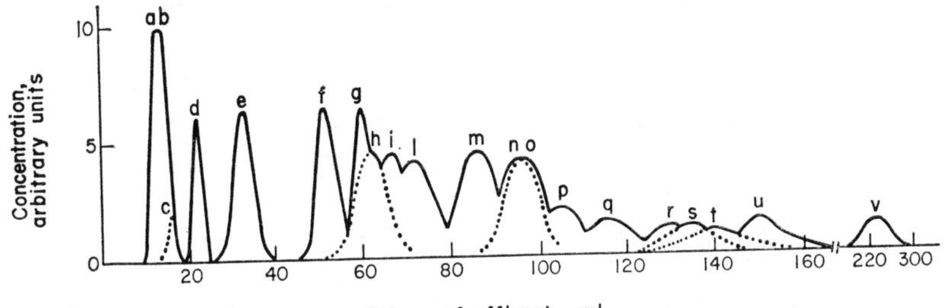

FIG. 2.10. Elution curves for aromatic amines on an alginic acid column with 1 M acetic acid as eluent. (a) Sulphanilic, methanilic and orthanilic acids; (b) *o*-arsanilic acid; (c) *o*-nitroaniline; (d) *p*-nitroaniline; (e) *p*-arsanilic acid; (f) 4-aminosalicylic acid; (g) *o*-aminobenzoic acid; (h) sulphanilamide; (i) *p*-aminoacetophenone; (l) 5-aminosalicylic acid; (m) *p*-aminobenzoic acid; (n) *o*-chloroaniline; (o) *p*-aminohippuric acid; (p) *m*-nitroaniline; (q) aniline; (r) *o*- and *p*-toluidine; (s) *m*-aminobenzoic acid; (t) *o*-anisidine and *o*-aminophenol; (u) *m*- and *p*-aminophenol; (v) α- and β-naphthylamine. (From L. Lepri *et al.*, *J. Chromatogr.* **49**, 239 (1970).)

layer when the R_f values are high, and greater when the R_f values are lower than 0·5 in 1 N acetic acid. In the latter case a column permits separations which are not foreseeable from the R_f, such as the separation of *o*-chloroaniline and *p*-aminobenzoic acid from a group of amines including toluidines, aminophenols, *o*-anisidine and *p*-aminobenzoic acid (cf. Table 2.28).

Considerable differences in the chromatographic results for the same amines when the column or thin layer techniques are applied have been remarked. In order to verify the existence of a relationship between the two sets of data, the equation

$$V_{\max} = V_{\text{int}} + \frac{A_1}{A_s} g\left(\frac{1}{R_f} - 1\right)$$

was applied, where:

V_{int} = interstitial volume of the column,

A_1/A_s = cross-sectional areas ratio of mobile and stationary phase on thin layer,

g = weight of the polymer in the column.[64, 76]

As from the data collected in Table 2.29, the above equation seems to fit for both alginic acid and carboxymethyl cellulose in the range of R_f

TABLE 2.29. VOLUME OF EFFLUENT RELATIVE TO THE PEAK OF THE ELUTION CURVE (V_{max}) AND R_f VALUE FOR SOME AROMATIC AMINES ON ALGINIC ACID WITH 1 M ACETIC ACID AS ELUENT

Compound used	V_{max}	R_f	$1/(R_f-1)$
o-Arsanilic acid	14·0	0·82	0·22
o-Nitroaniline	16·0	0·55	0·82
p-Nitroaniline	22·0	0·47	1·13
p-Arsanilic acid	32·5	0·41	1·44
4-Aminosalicylic-acid	52·0	0·26	2·85
o-Aminobenzoic-acid	60·5	0·23	3·35
Sulphanilamide	62·5	0·22	3·55
p-Aminoacetophenone	67·0	0·21	3·77
5-Aminosalicylic acid	71·5	0·20	4·00
p-Aminobenzoic acid	86·0	0·16	5·25

(From L. Lepri *et al.*, *J. Chromatogr.* **49**, 239 (1970).)

between 0·2 and 0·8 only. This was justified on the ground that for high R_f the solvent front influences the chromatographic behaviour of the compounds,[64] while for R_f lower than 0·2 a large experimental error is associated with the measurement.

Some interesting separations performed by column chromatography on alginic acid include the separation of 4-aminosalicylic acid from *m*-aminophenol, and the separations of the isomers of some aromatic amines, as presented in Table 2.30.

The resolution of racemic bases has been performed on both alginic acid and polygalacturonic acid, by column chromatography.[77, 78] The following bases were studied:

(±) threo $C_6H_5.CH(NH_2).CH(C_6H_5).COOCH_3$ I

(±) erythro $C_6H_5.CH(NH_2).CH(C_6H_5).COOCH_3$ II

(±) threo [piperidin-2-yl, N]–$CH(C_6H_5).COOCH_3$ (ritaline) III

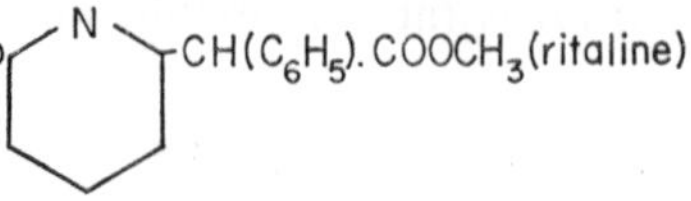

The bases were introduced into the column filled with the suitably swelled polymer, as diluted solutions in methanol–ether mixture 1 : 1. As presented

TABLE 2.30. SEPARATIONS ON AN ALGINIC ACID COLUMN WITH 1 M ACETIC ACID AS ELUENT

Compound used	Weight of compound placed on column (μg)	Volume range of the eluate (ml)	Recovery of base (%)*
o-Nitroaniline	10	13–19	93±4
p-Nitroaniline	30	19–25	95±3
m-Nitroaniline	30	95–115	92±4
o-Aminobenzoic acid	50	54–68	94±3
p-Aminobenzoic acid	50	74–98	94±3
m-Aminobenzoic acid	50	120–150	91±4
o-Arsanilic acid	50	10–18	95±3
p-Arsanilic acid	50	25–40	94±3
p-Arsanilic acid	100	22–45	96±3
m-Aminophenol	50	140–165	90±5
4-Aminosalicylic acid	50	42–60	93±4
4-Aminosalicylic acid	1000	32–75	97±2
4-Aminosalicylic acid	5000	26–90	98±2
5-Aminosalicylic acid	50	60–80	94±4

* The reported data are the means of several determinations.

(From L. Lepri, *J. Chromatogr.* **49**, 239 (1970).)

in Table 2.31, generally five aliquots of the eluate were collected, and the column was afterwards sectioned in three portions which were treated with hydrochloric acid solution 1% in methanol. The solutions were then neutralized with sodium bicarbonate, and the solvents were evaporated. The degree of swelling was found to be very important for the reproducibility of the separation, and the best was found in the range of 10–25 ml g^{-1}. In order to ensure a complete reaction of the amino group with the carboxyl group of the polymers, an excessive swelling is not strictly necessary, but it seems that the polymers through swelling reach the desired molecular asymmetry, thereby revealing the different behaviours of the two antipodes. The bases I and III (threo) behave identically on both polygalacturonic and alginic acids. The base II behaves differently towards the

TABLE 2.31. INFLUENCE OF THE DEGREE OF SWELLING ON THE SEPARATION OF THE RACEMIC BASE II ($\frac{1}{2}$ g) ON ALGINIC ACID COLUMNS (I.D. = 1·8 cm)

Fract. No.	Degree of swelling, ml g^{-1} 1·5		14·3		19·2		22·0		32·0	
	g	$[\alpha]_D^{20}$	g	$[\alpha]_D^{20}$	g	$[\alpha]_D^{20}$	g	$[\alpha]_D^{20}$	g	$[\alpha]_D^{20}$
1	0·002	—1·3	0·003	— 8·0	0·010	— 3·4	0·011	—15·4	0·041	—12·1
2	0·099	—0·8	0·008	—16·5	0·005	—20·0	0·058	—23·0	0·203	— 6·0
3	0·022	—0·8	0·026	—26·8	0·025	—27·7	0·059	—14·2	0·048	— 0·2
4	0·004	0	0·027	—17·3	0·034	—22·9	0·031	— 8·1	0·029	+ 3·1
5	0·004	0	0·023	—10·8	0·023	—17·0	0·020	— 5·2	0·020	+ 4·9
	0·131†		0·087†		0·097†		0·179†		0·341†	
6*	0·091	—0·3	0·105	— 0·3	0·068	— 2·1	0·105	+ 4·5	0·060	+ 9·1
7*	0·100	—0·3	0·140	+ 6·1	0·128	— 6·1	0·070	+ 7·7	0·030	+11·0
8*	0·090	+0·5	0·051	+ 8·0	0·114	+ 9·4	0·087	+11·6	0·022	+10·6

* Taken from the column after completion of the chromatographic process.
† Total of the basic substances in the eluate fractions.

(From C. Kratchanov *et al.*, *J. Chromatogr.* **43**, 66 (1969).)

two polymers. When carrying out a chromatographic separation on polygalacturonic acid the (+) antipode is eluted first, while on alginic acid the (—) antipode passes first. A systematic survey of the chromatographic behaviour of racemic bases on alginic acid and polygalacturonic acid was proposed by Kratchanov *et al.*, in view of identification purposes.[78]

DERIVATIVES OF ALGINIC ACID

For the alginic acid molecule two possibilities of obtaining esters exist: the carboxyl groups can be esterified or the hydroxyl groups.

Carboxyl groups esterified

The methyl ester was prepared by treatment of alginic acid with diazomethane and with hydrogen chloride in methanol[79, 80] and by the reaction of dimethyl sulphate with sodium alginate suspended in a non-aqueous liquid.[81] Esters can be formed under mild conditions, with 1,2-alkylene oxides;[82, 83] with propylene oxide the propylene glycol ester can be prepared for addition to foods, like jellies which are made by reaction with calcium salts. No information is available on the possible interactions of these derivatives with transition elements.

Hydroxyl groups esterified

The acetylation of alginic acid is an important chemical operation which has been studied by many authors. The best method which involves little degradation has been published by Schweiger,[69] and it is recalled here as it belongs to that class of procedures for changing properties of the polysaccharides, like starch and cellulose, whose acetates have great commercial value. In the preceding presentation of chromatographic data, alginic acid acetate has been recalled in several instances.

The usual methods cannot be applied for the acetylation of alginic acid; they include the reaction of acetyl chloride in the presence of an organic base, and the catalytic reaction with acetic anhydride or acetic acid. It is essential for a successful reaction to use a wet alginic acid fibre as starting material with as low an ash content as possible. A partial dehydration should be carried out in such a way that the residual water is left in an even distribution. Drying at elevated temperature is to be avoided as it brings about inactivation of alginic acid. In fact some water molecules associated with the hydroxyl groups of the polymer prevent an extended formation of hydrogen bridge bonding: when hydrogen bridge formation is allowed to take place, the substitution of the hydroxyl groups of interest becomes very difficult.

Products with degree of acetylation of up to 1·85 are obtained if the reaction is carried out at a moderate temperature. The theoretical value of two can be reached provided that the polymer is sufficiently pure and that the reaction temperature is raised considerably: however, in the latter case an extended degradation takes place.

For the preparation of ammonium diacetyl alginate, 1 kg of wet alginic acid (up to 80% water and 1% ash as calcium oxide) is washed with 1 l of glacial acetic acid in a mixer, and then is filtered on a Buchner filter, twice. The washed polymer is then placed in the mixer again with 1·5 l of glacial acetic acid and 1·5 l of acetic anhydride. While mixing, 2 ml of perchloric acid are added cautiously so that the temperature does not exceed 40–45 °C: this requires about 2 hr. After a total reaction time of 3 hr, the mixture was pressed out, and washed with distilled water to pH = 6·0.

The free diacetyl alginic acid is neutralized with ammonium hydroxide after suspending it in water, and the syrup so obtained is then poured into acetone and mixed thoroughly. The filtered product can be dried in air; the yield is 300–350 g with a degree of acetylation in the range 1·75–1·85.

The ammonium diacetyl alginate is soluble in water, insoluble in organic solvents, and swells in lower alcohols. It forms gels or precipitates with copper(II), tin(II), and with tri- and tetravalent ions. It does not precipitate or gelatinize with calcium, barium, magnesium, iron(II), manganese(II), or zinc.[83]

The gradual replacement of hydroxyl groups by acetyl groups is a viscosity reducing process, assuming that hydroxyl groups contribute to viscosity by their ability to associate with water molecules. Experimentally, the viscosity maximum is at a degree of acetylation lower than one.

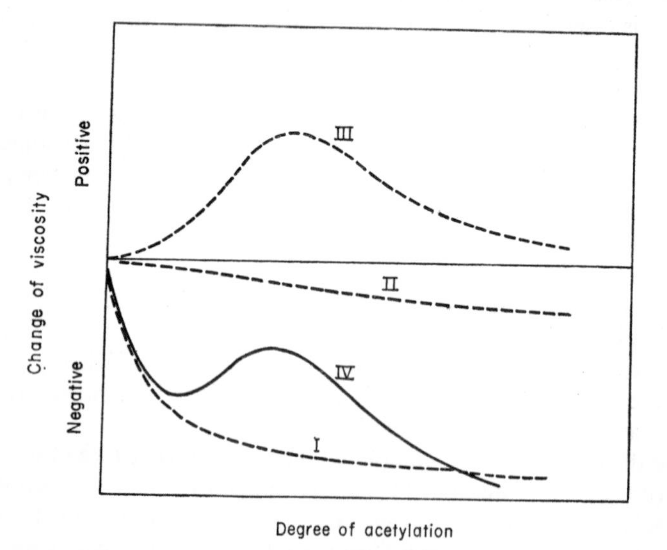

FIG. 2.11. Influence on viscosity by (I) hydrolysis, (II) introduction of acetyl groups, and (III) by reducing the extent of hydrogen bridge bondings. Curve (IV) results when (I), (II), and (III) are added. (From R. G. Schweiger, *J. Org. Chem.* **27**, 1786 (1962).)

If hydrogen bridge bondings between the vicinal hydroxyl groups block them to a certain extent, the viscosity maximum may correspond with a maximum of the number of hydroxyl groups. Since the hydroxyl groups of the monoacetate unit cannot form such a bond, the number of unblocked groups may increase during the first phase of the reaction, until the second step leading to the diacetate unit predominates. The viscosity change by this effect would be positive, reaching a maximum around D.A. = 1; then it would become negative and drop to a constant value, as shown in Fig. 2.11, curve III. The sum of the three curves in Fig. 2.11 yields the experimental curve. These results produce evidence of the presence of hydrogen bridge bondings between vicinal hydroxyl groups, as the reason for the low reactivity of alginic acid and alginates.

Alginic acid sulphates have been obtained by reaction with chlorosulphonic acid and pyridine as a catalyst.[84] In this field the interest is centred on their anticoagulant activity.

Other esters of alginic acid have been reported and investigated in view of commercial applications.

Ethers

Methyl ethers have been used in research on the constitution of alginic acid. Carboxymethylalginic acid can be obtained as the sodium salt, from sodium alginate with chloroacetic acid in the presence of sodium hydroxide,[85] following the same procedure used for carboxymethylcellulose. Henkel[86] has reported an amide of alginic acid.

Salts of organic bases

Only a few organic bases have been studied in the salt formation of alginic acid. The principal aim was to affect the solubility of alginic acid in organic solvents. Tributylamine, phenyltrimethylammonium and benzyltrimethylammonium alginates are soluble in absolute ethanol.[87] Triethanolamine alginate is soluble in 75% aqueous ethanol. The alginate of Primene 81 R is soluble in polar organic solvents, but it is precipitated upon addition of water.[88] It is known that quaternary ammonium compounds with long hydrocarbon chains as acetyltrimethylammonium bromide, react with alginic acid to form a precipitate.[89]

REFERENCES

1. E. Percival and R. H. McDowell, *Chemistry and Enzymology of Marine Algal Polysaccharides*, Academic Press, London, 1967.
2. E. Cerma, *Boll. Soc. Adr. Sci., Trieste* **56,** 97 (1968).
3. A. B. Steiner and W. H. McNeely, *Ind. Engng Chem.* **43,** 2073 (1951).
4. S. Miklestad, Norway Patent 111,426 (1967).
5. E. L. Hirst, E. Percival and J. K. Wold, *J. Chem. Soc.* 1493 (1964).
6. E. L. Hirst, J. K. N. Jones and W. O. Jones, *J. Chem. Soc.* 1880 (1939).
7. S. K. Chanda, E. L. Hirst, E. G. V. Percival and A. G. Ross, *J. Chem. Soc.* 1833 (1952).
8. H. J. Lucas and W. T. Stewart, *J. Am. Chem. Soc.* **62,** 1070 (1940).
9. H. J. Lucas and W. T. Stewart, *J. Am. Chem. Soc.* **62,** 1792 (1940).
10. F. G. Fischer and H. Doerfel, *Hoppe-Seyler's Z. Physiol. Chem.* **301,** 224 (1955).
11. F. G. Fischer and H. Doerfel, *Hoppe-Seyler's Z. Physiol. Chem.* **302,** 186 (1955).
12. D. W. Drummond, E. L. Hirst and E. Percival, *Chem. Ind.* 1088 (1958).
13. D. W. Drummond, E. L. Hirst and E. Percival, *J. Chem. Soc.* 1208 (1962).
14. R. L. Whistler and R. Schweiger, *J. Am. Chem. Soc.* **80,** 5701 (1958).
15. R. L. Whistler and J. N. BeMiller, *J. Am. Chem. Soc.* **82,** 457 (1960).
16. E. L. Hirst and D. A. Rees, *J. Chem. Soc.*, 1182 (1965).
17. A. Haug, B. Larsen and O. Smidsrød, *Acta Chem. Scand.* **20,** 183 (1966).
18. H. Kingstad and G. Lunde, *Kolloidzeitschrift* **83,** 202 (1938).
19. W. T. Astbury, *Nature, Lond.* **155,** 667 (1945).
20. K. J. Palmer and M. B. Hartzog, *J. Am. Chem. Soc.* **67,** 1865 (1945).
21. E. Tallis, *J. Text. Inst.* **41,** T-151 (1950).
22. C. Sterling, *Biochim. Biophys. Acta* **26,** 186 (1957).
23. J. O. Warwicker, *Shirley Inst. Mem.* **31,** 41 (1958).
24. E. Frei and R. D. Preston, *Nature, Lond.* **196,** 130 (1962).
25. E. G. V. Percival and E. Percival, *Structural Carbohydrate Chemistry*. J. Garnet Miller, London, 1962.
26. D. Cozzi, P. G. Desideri, L. Lepri and G. Ciantelli, *J. Chromatogr.* **35,** 396 (1968).
27. F. G. Donnan and R. C. Rose, *Can. J. Res.* B-**28,** 105 (1950).
28. W. H. Cook and D. B. Smith, *Can. J. Biochem. Physiol.* **32,** 227 (1954).
29. A. Haug, *Acta Chem. Scand.* **15,** 950 (1961).
30. A. Haug, *Acta Chem. Scand.* **15,** 1794 (1961).
31. N. H. Chamberlain, A. Johnson and B. Speakman, *J. Soc. Dyers Colour.* **61,** 13 (1945).
32. M. C. Cameron, A. G. Ross and E. G. V. Percival, *J. Soc. Chem. Ind.* **67,** 161 (1948).
33. A. S. Perlin, *Can. J. Chem.* **30,** 278 (1952).
34. A. Jensen, I. Sunde and A. Haug, *The Quantitative Determination of Alginic Acid.* Rept. N. 12. Norwegian Inst. Seaweed Res. Trondheim.
35. E. G. Brown and T. J. Hayes, *Analyst* **77,** 445 (1952).
36. A. Haug and B. Larsen, *Acta Chem. Scand.* **16,** 1908 (1962).

37. E. A. BALAZS, K. O. BERNTSEN, J. KAROSSA and D. A. SWAN, *Analyt. Biochem* **12,** 547 (1964).
38. A. HAUG and O. SMIDSRØD, *Acta Chem. Scand.* **19,** 341 (1965).
39. J. T. TRIFFITT, *Nature, Lond.* **217,** 457 (1968).
40. O. SMIDSRØD and A. HAUG, *Acta Chem. Scand.* **19,** 329 (1965).
41. A. HAUG, *Acta Chem. Scand.* **13,** 1250 (1959).
42. H. THIELE and G. ANDERSEN, *Kolloidzeitschrift* **140,** 76 (1955).
43. A. HAUG, B. LARSEN and O. SMIDSRØD, *Acta Chem. Scand.* **20,** 183 (1966).
44. S. MYKLESTAD and A. HAUG, *Proc. 5th Int. Symp. Halifax*, Nova Scotia (YOUNG and MCLACHLAN, Eds.), Pergamon Press (1965).
45. O. VAN DER BORGHT, S. VAN PUYMBROECK and J. COLARD, *Health Phys.* **21,** 181 (1971).
46. S. SKORYNA, T. PAUL and D. WALDRON-EDWARD, *Can. Med. Ass. J.* **91,** 285 (1964).
47. G. HARRISON, E. HUMPHREYS, A. SUTTON and H. SHEPERD, *Science* **152,** 655 (1966).
48. G. PATRICK, T. CARR and E. HUMPHREYS, *Int. J. Radiat. Biol.* **12,** 427 (1967).
49. K. KOSTIAL, T. MALJKOVIC, M. KADIC, R. MANITASEVIC and G. HARRISON, *Nature* **215,** 182 (1967).
50. J. F. STARA and D. WALDRON-EDWARD, *Proc. Symp. Diagnosis Treatment of Deposited Radionuclides* (H. A. KORNBERG, Ed.), p. 340 (1968).
51. L. B. COLVIN, G. R. GREGER and J. R. COUGH, *Proc. Soc. Exp. Biol. Med.* **124,** 566 (1967).
52. L. MILLIN and J. J. B. ANDERSON, *J. Nutr.* **97,** 181 (1969).
53. E. HUMPHREYS and J. TRIFFITT, *Nature, Lond.* **219,** 1172 (1968).
54. O. VAN DER BORGHT, J. COLARD, S. VAN PUYMBROEK and R. KIRCHMANN, Alginic acid, in: *Radioecological Concentration Processes* (B. ABERG and E. P. HUNGATE, Eds.), Pergamon Press, Oxford, 1967, p. 589.
55. G. PATRICK, *Nature, Lond.* **216,** 815 (1967).
56. W. MOORE and R. ELDER, *Nature, Lond.* **206,** 815 (1967).
57. R. HESP and B. RAMSBOTTOM, PG Report 686, UKAEA (1965).
58. A. SUTTON, *Nature, Lond.* **216,** 1005 (1967).
59. A. HODGKINSON, B. NORDIN and O. B. C. HAMBLETON, *Can. Med. Ass. J.* **97,** 1139 (1967).
60. T. CARR, G. HARRISON, E. HUMPHREYS and A. SUTTON, *Int. J. Radiat. Biol.* **14,** 225 (1968).
61. O. VAN DER BORGHT, J. COLARD and R. BOULENGER, *Health Physics* **23,** 240 (1972).
62. G. RASPI and M. G. CAMICI ZAPPELLI, *Ann. Chim.* **58,** 922 (1968).
63. D. COZZI, P. G. DESIDERI, L. LEPRI and G. CIANTELLI, *J. Chromatogr.* **35,** 396 (1968).
64. M. LEDERER, *Bull. Chem. Soc. France* **16** (1966).
65. D. COZZI, P. G. DESIDERI and L. LEPRI, *J. Chromatogr.* **40,** 130 (1969).
66. D. COZZI, P. G. DESIDERI, L. LEPRI, *J. Chromatogr.* **35,** 405 (1968).
67. D. COZZI, P. G. DESIDERI and L. LEPRI, *J. Chromatogr.* **42,** 532 (1969).
68. M. Y. DOLMATOVA and A. P. PANTELEEVA, *Radiokhimiya* **10,** 379 (1968).
69. R. G. SCHWEIGER, *J. Org. Chem.* **27,** 1786 (1962).

70. R. G. SCHWEIGER, *J. Org. Chem.* **27,** 1789 (1962).
71. E. KOROS, ZS. REMPÓRTH-HORVÁTH, A. LASZTITY and E. SCHULEK, *Proc. Symp. Soc. Anal. Chem.*, Nottingham, 1965, Heffer, Cambridge, p. 593 (1965).
72. R. G. SCHWEIGER, *Kolloidzeitschrift* **196,** 47 (1964).
73. D. COZZI, P. G. DESIDERI, L. LEPRI and V. COAS, *J. Chromatogr.* **40,** 138 (1969).
74. L. LEPRI, P. G. DESIDERI and V. COAS, *J. Chromatogr.* **64,** 271 (1972).
75. D. COZZI, P. G. DESIDERI, L. LEPRI and V. COAS, *J. Chromatogr.* **43,** 463 (1969).
76. L. LEPRI, P. G. DESIDERI, V. COAS and D. COZZI, *J. Chromatogr.* **49,** 239 (1970).
77. C. KRATCHANOV and M. POPOVA, *J. Chromatogr.* **37,** 297 (1968).
78. C. KRATCHANOV, M. POPOVA, T. OBRÉTÉNOV and N. IVANOV, *J. Chromatogr.* **43,** 66 (1969).
79. H. J. LUCAS and W. T. STEWART, *J. Am. Chem. Soc.* **62,** 1792 (1940).
80. E. F. JANSEN and R. JANG, *J. Am. Chem. Soc.* **68,** 1475 (1946).
81. W. H. MCNEELY and J. J. O'CONNELL, U.S. Patent 2,860,130 (1958).
82. A. B. STEINER and W. H. MCNEELY, *Ind. Engng Chem.* **43,** 2073 (1951).
83. A. B. STEINER and W. H. MCNEELY, *Am. Chem. Soc. Advances Chem. Series,* No. 11, 72 (1954).
84. E. G. SNYDER, Brit. Pat. 676, 564 (1950).
85. W. H. MCNEELY and J. J. O'CONNELL, U.S. Patent 2,902,479 (1959).
86. HENKEL and GmBH Co., Brit. Patent 768,309 (1957).
87. W. H. MCNEELY, U.S. Patent 2,688,598 (1954).
88. J. L. BOYLE, Brit. Patent 835,009 (1960).
89. J. E. SCOTT, *Methods of Carbohydrate Chemistry* (R. C. WHISTLER, Ed.), Vol. V, page 38, Academic Press, London, 1965.

CHAPTER 3

CHITIN

PRESENTATION

Chitin is a polysacharide constituted of β–(1 → 4) 2-acetamido-2-deoxy-D-glucose units, some of them being deacetylated. This natural polymer that can be called poly-N-acetyl-D-glucosamine, can be formally considered a derivative of cellulose where the C–2 hydroxyl groups have been completely replaced by acetamido groups.

$$\left[-O- (\text{CH}_2\text{OH}, \text{OH}, \text{NHCOCH}_3 \text{ pyranose}) -O- (\text{NHCOCH}_3, \text{OH}, \text{CH}_2\text{OH} \text{ pyranose}) - \right]_n$$

The interest in this polymer from the chromatographic point of view has only recently been aroused: as the nitrogen per cent of chitin is theoretically 6·89 against about 1·25 or less for artificially substituted celluloses, chitin was expected to yield better performances than substituted celluloses in inorganic chromatography, as it can act as a chelating solid agent.[1]

OCCURRENCE AND AVAILABILITY OF CHITIN

It would seem that the *in vitro* formation of a polysaccharide represents a simpler problem than, for instance the synthesis of a nucleic acid or a protein, in which a complicate array of components has to be ordered in a specific sequence. Nevertheless, although with enormous advances in the area of homopolysaccharides and heteropolysaccharides synthesis, we lack information about the macromolecular aspects of the biosynthesis of polysaccharides, and at the present time we cannot think of a convenient production of a polymer like chitin by biosynthesis.[2, 3]

Therefore it is necessary to survey and present the occurrence in nature of chitin, and to discuss its availability.

Chitin occurs widely in lower animals, fungi, etc. The exoskeletons of crabs and lobsters have long since attracted attention as a source of raw material for chitin production as the dry arthropod exoskeletons contain from 20 to 50% chitin. Very little is known about the metabolism of chitin in animals, but it is clear that it acts as a carbohydrate and nitrogen reserve and it is associated with proteins to form glycoproteins. The supporting or protecting function of chitin seems to be of minor importance.

Relatively recent reviews [4–6] have covered the subject of the arthropod cuticle structure, which can be considered a two layer structure of chitin protein complexes, with the exocuticle hardened by polymerization of the proteins with polyphenols, and admixed with calcium salts. Chitin can be found in gut linings, tracheae, wing coverings, and the other parts of the body of the lower animals, as described by Richard, however, as far as the cuticle is concerned, chitin is accompanied, besides proteins, by an important inorganic fraction, mostly constituted by calcium carbonate and phosphate.[4, 7, 8] In Table 3.1 one can remark that the cuticle of the edible crab is the one that contains the highest proportion of chitin to protein.

Crab cuticle are abundantly available. *Carcinus maenas* and other crabs are currently cultivated in many places in the world. particularly at Chioggia, Italy, for instance, where the yearly production amounts to about 350 tons. The crabs are collected during the moulting periods (April and September) and their cuticles can be dried and chemically treated.[9] Factories for canning crab or shrimp meat or similar, particularly in the Scandinavian countries and Japan, discard daily very large amounts of chitinous raw material.

The moulted cuticle contains more inorganic than organic constituents; calcium carbonate being the preponderant fraction. Carbonate can be eliminated by an acid treatment; a tricholoroacetic acid 10% solution is

TABLE 3.1. PROPORTIONS OF ORGANIC COMPONENTS IN ARTHROPOD CUTICLES

Source	Proportion (% of organic fraction; dry weight)	
	Chitin	Protein
Arachnida		
Buthus (scorpion)	31·9	68·1
Mygale (spider)	38·2	61·8
Insecta		
Locusta, clytra and wings	23·7	76·3
Periplaneta (cockroach) av.	35·0	—
Coleoptera		
Dytiscus (water beetle)		
adult elytra	37·4	62·6
Lepidoptera		
Bombyx (silk worm), larvae	44·2	55·8
Crustacea (Decapodus)		
Cancer (edible crab)	71·4	13·3
Eupagurus (hermit crab)		
calcified	69·0	31·0
non-calcified	48·2	51·8

(From J. S. Brimacombe *et al.*, *Mucopolysaccharides*, Elsevier.)

suitable for selective elimination of the inorganic part, and for the determination of the total organic part (protein + chitin) which is called "artropodin complex"; in this complex glutamic and aspartic acids and the basic amino acids histidine, lysine, and arginine are present in varying proportions in various species, and sometimes the relative concentration of amino acids in the cuticle has been reported to vary throughout the intermoult period.[10, 11]

There is a considerable amount of evidence to support the indication that in native chitin not all the glucosamine residues are acetylated, and the free amino groups may accept a proton and carry a positive charge. Giles[12] performed elemental analysis on lobster chitin which had been very gently deproteinized to avoid artificial deacetylation, and concluded that the best estimated composition by weight was: *N*-acetylglucosamine, 82·5%; glucosamine, 12·5% and water 5·0%. Waterhouse *et al.* extracted glucosamine and *N*-acetylglucosamine from crab cuticle with a chitinase preparation which showed no deacetylase activity, and found up to 10% glucosamine in native chitin. This corresponds to one deacetylated residue out of six on the polymer chain. This point is further discussed in the light of the macrostructure of this polymer.

Marchessault[13] showed that alkaline purification of chitin led to a change in pH to 3·5, as more and more free amino groups were becoming available and collected protons to form $—NH_3^+$.

It is suggested that the degree of deacetylation and subsequent presence of charge is significant in determining the orientation of the polymer when the ionic composition of the deposition zone is varied by the cells. One should also remember that cellulose in plants contains a small number of uronic acid residues. Orientation control is thus effected by both pH changes in the extracellular environment, and ionic changes. A rhythmical ion pump could control lamellogenesi frequency. Divalent ions would have a greater screening effect on polar groups than monovalent ions.[14] Chitin free oothecae of cockroaches that contain large amounts of Ca^{++} lack protein orientation. Scudamore[15] has shown than inorganic ions have biological significance on the daily orientation changes in insect cuticle. Travis[16] expressed the opinion that the continued increase in mineral content, which takes place with the mineralization of crustacean cuticle and gastroliths, is due to the epidermal cells control of calcium, carbonate, phosphate and pH changes in the extracellular environment.

The dry cuticle nitrogen content is 1·81% while the dry soft crab after moulting contains 9·37% nitrogen. For a 10 g crab, the dry cuticle weighs 4·75 g, corresponding to 86 mg nitrogen which represents 15% of the total nitrogen that is lost by the crab on moulting. (See also Table 3.2.[17])

Chitin sediments in sea-water were estimated to amount to several billion tons per year, mostly due to moulted copepod exoskeletons.[18] Chitin is therefore, largely present in the ocean. In sediments, chitin affects the C/N ratio which reaches sometimes value 10, and which is controlled by chitinolytic bacteria.

Copepods are the most abundant multicellular animals in the world, and they produce eleven cast exoskeletons for every adult; these exuvia may be capable of taking up elements during their slow sinking.

TABLE 3.2. COMPOSITION OF THE ORGANIC MATTER OF THE CRUSTACEAN TEGUMENT *in toto*

	Decapod							Isopods	Cirripedes	Branchiopods
	Natantia		*Reptantia*							
	Palaemon serratus		*Nephrops norvegicus*		*Maia squinado*	*Eupagurus Prideauxi*		*Ligia oceanica*	*Lepas anatifera*	*Triops cancriformis*
	Cephalothorax	Abdomen	Thorax	Abdomen	Cephalothorax-dorsal	Pincets	Abdomen	Tergites	Peduncle	Mouth part
Total nitrogen	P. 100.	P.100.	P. 100.	P. 100.	P. 100.	P. 100.	P. 100.	P. 100.	P. 100.	P. 100.
	9·67	9·58	7·74	8·03	8·50	8·7	8·9	7·44	9·57	9·8
Chitin	64·5	60·5	77·5	76·3	72·5	70·3	63·2	78·5	58·3	61·4
Nitrogen content of chitin	6·59	6·6	6·6	6·66	6·6	70·3	63·2	6·7	6·6	61·4
Chitin N/total N	43·9	41·6	66·0	63·2	56·2	70·3	63·2	70·6	40·2	61·4
Aminic nitrogen	2·25	2·44	2·2	2·16	2·62	70·3	63·2	1·53	3·06	61·4
Protein	14·0	15·2	13·7	13·7	16·3	70·3	63·2	9·5	19·0	61·4
Protein/chitin	21·7	25·1	17·8	17·9	22·4	70·3	63·2	12·1	32·7	61·4

(From M. Lafon, *Ann. Sci. Nat. Ser. Bot. Zool.* **11**, 113 (1943).)

TABLE 3.3. ELEMENTAL CONCENTRATIONS IN THE ZOOPLANKTON. APPROXIMATE CONCENTRATIONS OF ELEMENTS PER GRAM WET WEIGHT OR DRY WEIGHT CAN BE OBTAINED BY DIVIDING THE ASH WEIGHT VALUES BY 25 (WET WT: ASH WT) OR BY 3 (DRY WT: ASH WT)

Sample No.	Depth (m)	Pb	Zn	Fe	Cd	Co	Cu	Ni	Mn	Sr	Ca	Mg
		(μg:g ash)										
1	surface	110	580	3,600	21	60	200	185	—	780	85,000	60,000
2	surface	85	1,200	3,800	21	60	110	230	—	1,060	80,000	55,000
3	surface	95	750	4,400	23	60	270	125	—	1,560	190,000	52,000
4	surface	25	550	2,450	21	40	145	85	—	1,280	165,000	48,000
5	surface	370	430	2,000	27	40	215	50	—	1,500	105,000	54,000
6	surface	ND*	360	3,000	16	30	40	85	—	820	75,000	42,000
7	surface	25	360	1,700	10	45	40	65	70	350	110,000	33,000
8	surface	35	1,120	3,500	9	50	30	55	70	940	42,000	38,000
9	surface	280	800	2,900	7	25	50	75	ND	170	97,000	35,000
10	surface	120	420	1,750	10	30	50	45	ND	440	90,000	40,000
11	100	—	2,700	5,400	—	20	620†	1,070†	—	—	85,000	—
12	100	140	950	5,000	—	40	190	340	—	920	85,000	48,000
13	100	90	1,200	3,800	12	40	120	205	—	1,140	53,000	62,000
14	100	75	1,900	4,800	—	30	550†	135	—	2,100	62,000	60,000
15	100	100	1,250	5,600	18	40	220	220	—	920	58,000	60,000
16	100	410	1,600	3,400	36	50	110	85	—	2,060	175,000	50,000
17	100	150	770	3,900	—	—	180	240	—	2,400	62,000	31,000
18	100	130	840	3,000	8	45	50	150	70	290	50,000	42,000
19	100	3,200†	2,300	3,600	13	30	60	80	90	650	165,000	39,000
20	150	3,200†	2,200	4,400	11	35	60	60	100	900	170,000	43,500
21	200	330	2,400	3,200	13	35	240	75	80	380	80,000	43,500
22	500	225	3,600	4,200	12	45	90	65	100	740	215,000	29,500
Avg	surface	117	657	2,900	16	44	115	100	<70	890	103,900	45,700
Avg	>99 m	183	1,809	4,200	15	37	132	150	88	1,140	105,000	46,200
Avg	all samples	148	1,285	3,600	16	40	123	126	<83	1,020	104,500	46,000

* ND = not detected.
† These values were not included in the means.
(From J. H. Martin, *Limnol. Oceanogr.* **15**, 756 (1970).)

As can be seen from the Table 3.3, zooplankton can be of great importance in the cycling of elements in the world's oceans.[19] They are second only to phytoplankton in abundance. They can transport elements in various ways; by vertical migration across mixing barriers, by incorporation of metals into faecal products that sink rapidly, by the passage of elements to the trophic levels, and, what is most important, by the moulting of exoskeletons mentioned above and by sinking of skeletal structures after death. Studies on the interaction of these chitinous materials and metal ions or complexes are extremely limited.

Table 3.3 shows that Cd and Co are slightly higher in the surface samples, while Cu, Fe, Mn, Ni, Pb, Sr, and Zn concentrations are higher in the deep samples. Szabo reported[20, 21] higher Cr, Fe, and Ti concentrations in deep samples, as well as Cu, Mn and Ni. Zooplankton concentration factors were calculated by converting average elemental concentrations for all samples analyzed in Table 3.3 from an ash to a wet weight basis; these values were then divided by the amount of the

TABLE 3.4. CONCENTRATION FACTORS FOR THE ZOOPLANKTON

Element	μg/g wet zooplankton	μg/g sea-water	Concn factor
Pb	5·9	0·00003	197,000
Fe	144·0	0·01	14,400
Cd	0·6	0·0001	6,000
Zn	51·0	0·01	5,100
Co	1·6	0·0005	3,200
Ni	5·0	0·002	2,500
Mn	3·3	0·002	1,650
Cu	4·9	0·003	1,630
Ca	4200·0	400·0	10·5
Sr	40·8	8·0	5·1
Mg	1800·0	1350·0	1·3

(From J. H. Martin, *Limnol. Oceanogr.* **15,** 756 (1970).)

given element found in an equal weight of sea-water. Listed as they are in Table 3.4, the metal concentration factors are in support of the Bowen's statement that "among cations (including metallic elements as Fe which may exist as colloids in the sea) the order of affinity for living matter is, broadly speaking: tetravalent and trivalent elements > divalent transition metals > divalent Group II A metals > Group I metals".[22] Concentration factors in living matter are low for Mg, Sr, and Ca. Some comments are due for the lead figure 197,000. The concentration of dissolved lead in deep ocean water should be 0·03 $\mu g\ l^{-1}$, but surface waters contaminated with the lead fall-out from engine exhausts can contain ten to twenty times as much lead. The use of 0·4 for the calculation of the concentration factor yields 14,750 which is still the highest figure, nearly like iron. In spite of no known biological functions, lead is therefore highly concentrated, and this fact can be compared to similar findings for other metals in planktonic organisms and brown algae.

Tatsumoto and Patterson[23] have estimated that zooplankton can remove 2×10^{11} g of lead per year, and this means one half of the amount yearly introduced in the oceans. While the collection action by living matter is quite important, the dead organisms or the products of their metabolism are also important. Much lead in fact can be carried to the sea bottom via copepod moults.

Some of the elements absorbed by the living organism once the organism is dead or once it has left its exoskeleton by moulting, may be replaced by other metals that form more stable complexes with the organic ligands, for instance copper, nickel and lead instead of strontium. The enormous amount of chitin which falls to the sea bottom collects elements by various mechanisms, adsorption–exchange–chelation, and in the light of the long time available for this process to occur, it is quite probable that this will prove to be a major factor in the transport of transition metal ions in the sea. Too little is known about the fate of the copepod exuvia after leaving the surface waters, but it might be well possible that the high metal concentration recorded at depth around 700 m in tropical seas is related to a release from chitin consequent to its degradation, in connection with a retardation in its sinking due to physical reasons (temperature gradients).[24, 25]

In any case it is certain that chitin with collected metals reaches the sea bottom in large amounts.[26] Skadovskii found accumulated residues of chitinous cuticles of Crustacea characteristic of zooplankton, chiefly daphnia, in oozes. Chitinous skeletons are characteristic of fossil forms of trilobites and graptolites.[27] Deposits of graptolites found in Dictyanema shales, indicate the essential role of chitin in formation of the organic mass of shales. Deposits encountered in old rocks (cenozoic)

TABLE 3.5. DISTRIBUTION OF CHITIN IN FUNGI*

Class	Order	Chitin	References
Phycomycetes	Chytridiales	+	von Wettstein, 1921; Nabel, 1939; Ajello, 1948
	Blastocladiales	+	Nabel, 1939
	Monoblepharidales	–	von Wettstein, 1921
	Saprolegniales	–	von Wettstein, 1921; Frey, 1950
	Peronosporales	–	von Wettstein, 1921; Frey, 1950
	Mucorales	+	Heyn, 1936; van Iterson, Meyer, and Lotmar, 1936; Schmidt, 1936; Blumenthal and Roseman, 1957
Ascomycetes	Laboulbeniales	+	Richards, 1954
	Sphaeriales	+	Blumenthal and Roseman, 1957
	Aspergillales (Plectascales)	+	Blank, 1953
	Saccharomycetales	†	Frey, 1950; Houwink, Kreger, and Roelofsen, 1951; Roelofsen and Hoette, 1951; Kreger, 1954; Miller and Phaff, 1958; Shifrine and Phaff, 1958
Basidiomycetes	Polyporales	+	Gilson, 1895; Proskuriakow, 1926; Schmidt, 1936
	Agaricales	+	Gilson, 1895; Scholl, 1908; Gonell, 1926; Proskuriakow, 1926; Khouvine, 1932; Locquin, 1943; Blumenthal and Roseman, 1957
Fungi imperfecti	Moniliales	+	Sumi, 1928; Thomas, 1928; Khouvine, 1932; May and Ward, 1934; Schmidt, 1936; Castle, 1945; Smithies, 1952; Blank, 1953, 1954; Kreger, 1954; Ballio, Casinovi, and Serlupi-Crescenzi, 1956; Blumenthal and Roseman, 1957

+ Present in all species tested.
– Absent from all species tested.
* Classification of Bessey, 1950.
† Present in most families, but apparently absent from *Saccharomyces* and *Schizosaccharomyces*.
[From N. Sharon, in *The Amino Sugars* (E. A. Balazs and R. W. Jeanloz, Eds.), Academic Press, London, 1965.]

contained thin organic films having several per cent of chitin content up to 15%.[28-30]

The examples reported illustrate the wide and massive distribution of chitin in contemporary and ancient organisms. However, no large accumulation of fossil chitin in undegraded form has been reported to date. While in certain conditions chitin is quite stable (chitinous fungal filaments have been reported in coal and coaly shale, and chitin from insect wings was found in coal 25 million years old) chitin does not remain unchanged in natural processes even though its degradation takes a long time, therefore very large deposits of fossil chitin are not expected to be found.[31]

The distribution of chitin in fungi is presented in Table 3.5. In the plant kingdom the occurrence of chitin is limited to fungi and green algae.[32] All fungi, with few possible exceptions have chitinous cell walls. It should be remarked however that earlier research on this topic included

TABLE 3.6. PERCENTAGE OF HEXOSE AND AMINO SUGAR WHICH REMAINED IN THE INSOLUBLE MATERIAL AFTER THE ACTION OF CHITINASE AND ALKALI ON THE CELL WALL OF THE YEAST PHASE OF *Paracoccidioides Brasiliensis*

Treatment	Hexose*	Amino sugar*
	%	%
Chitinase		
3 days	92·0	41·6
7 days	91·5	41·0
14 days	86·8	14·4
NaOH, 7 hr	14·0	97·9
Control, 14 days†	99·2	100·0

* The percentage of the insoluble sugars was calculated from the amount of sugar in the supernatant fluids and sediments after treatment. Hexose was determined as glucose and amino sugar as glucosamine.

† Only buffer and toluene were used in the control.

(From L. M. Carbonell, *J. Bacteriol.* **101**, 636 (1970).)

identification of chitin by microchemical and colour tests, and confirmation would be welcome.

Chitin is present in green algae, but not in bacteria and actynomycetes; several authors carried out researches on yeasts (saccharomycetes). Using microchemical methods of detection, chitin was found in a variety of filamentous and non filamentous yeasts,[33, 34] but some findings were not supported by X-ray diffraction studies. Kreger[35] obtained chitin diffraction patterns from numerous yeasts, after purification with acids

TABLE 3.7. CHITIN CONTENT OF VARIOUS FUNGI

Organism	Growth temp.	4 days' growth			8 days' growth		
		Chitin	Dry wt	pH	Chitin	Dry wt	pH
	°C	%	mg/100 ml		%	mg/100 ml	
Alternaria species	20	11·5	37	5·6	10·1	338	2·8
Aphanomyces laevis	25	7·0	182	2·9	7·7	620	2·0
Aspergillus flavus	25	13·6	623	2·2	13·2	981	2·2
Aspergillus flavus	25	22·4	714	2·4	25·6	990	2·2
Aspergillus niger	25	8·8	1056	2·0	14·6	830	2·3
Aspergillus niger	25	13·8	332	2·0	11·7	696	2·5
Aspergillus niger	20	8·9	645	2·3	21·2	418	2·3
Aspergillus niger	20	7·2	743	2·4	14·8	725	2·3
Aspergillus niger	20	9·6	700	2·3	22·4	448	2·2
Aspergillus oryzae	25	18·6	577	2·1	10·4	1038	2·0
Aspergillus parasiticus	25	14·8	930	2·1	17·1	1009	2·1
Aspergillus parasiticus	20	14·4	592	2·8	26·2	845	2·3
Aspergillus parasiticus	25	15·7	795	2·3	24·9	594	2·1
Collybia species	20			5·8	3·1	116	5·2
Dactylium dendroides	20	6·1	31	5·6	9·4	144	3·6
Geotrichium species	20	2·1	116	4·8	10·4	181	2·7
Glomerella cingulata	20	3·8	92	3·5	6·3	424	2·8
Helminthosporium sativum	20	4·4	58	5·6	6·4	284	3·0
Mucor rhizopodiformis	20	6·7	58	4·9	7·9	598	2·6
Neurospora crassa	20	2·4	200	3·3	2·6	163	3·3
Neurospora tetrasperma	25	4·6	278	2·8	14·7	990	2·2
Penicillium notatum	20	16·1	655	2·4	24·9	716	2·2
Penicillium species	20	5·8	262	3·0	6·5	556	2·3
Rhizopus nigricans	25	8·0	25	5·5	12·2	28	5·6
Rhizopus species	20	5·3	23	5·6	3·1	506	2·5

(From H. J. Blumental *et al.*, *J. Bacteriol.* **74**, 222 (1957).)

and alkali. Carbonell[36] observed short and thick fibres on the outer surface of the yeast phase of *Paracoccidioides Brasiliensis*, and long and thin fibres on the inner surface. The long fibres disappear with chitinase and are chitin fibres; the short ones disappear under alkali treatment and are composed of α-glucan. Some results are shown in Table 3.6.

In the fungal cell wall, chitin does not exist alone, but it is associated with other compounds.[37] Purified chitin from the fleshy fungi generally is recovered at the 3 or 5% since most of the alkali-resistant material amounting up to 40% of the dry weight of the organism is not chitin and can be destroyed by permanganate oxidation. Using purification by acids, Kreger[35] isolated chitin from *Agaricus campestris* with a yield of 35%, on which identification was done by X-ray diffraction. The data obtained with twentyfive cultures are summarized in Table 3.7.[38]

The identity of fungal chitin with the animal chitin was clearly established by chemical[39] and enzymatic degradation,[40,41] infrared spectra,[42] X-ray diffraction pattern comparison,[43, 44] optical data,[45] and viscosity measurements.[46]

In spite of the said identity, data were published on the collection of uranium from solutions of uranyl sulphate by fungal and animal chi-

TABLE 3.8. COLLECTION OF URANIUM FROM URANYL SULPHATE SOLUTIONS ON FUNGI OR CRAB CHITIN

Sample	Chitin g	pH	Time hr	Uranium collected %	Uranium left in solution %
Chitin	0·5	4·0	4	60	40
from fungi	0·5	4·0	90	65	35
	0·5	4·0	90	65	35
Chitin	0·5	4·0	90	20	80
from crab	0·5	6–7	90	25	75
	1·0	6–7	90	40	60
	0·5	8·0	90	60	40

(From S. M. Manskaya *et al.*, *Geochemistry of Organic Substances*, Pergamon Press, 1968.)

tins;[26] these data, presented in Table 3.8, indicate that from the point of view of the metal ion collection ability or capacity chitins are not equivalent: in fact, while fungal chitin removes 65% of the uranium used for the experiment, chitin from shells under the same conditions removes only 20%. Uranium forms a stable combination with chitin and Manskaya reported a 55% elution yield with 5% ammonium carbonate, and a 25% yield with 0·1 N oxalic acid. Data obtained by the present author on uranium collection on chitin have been published.[1]

DETECTION OF CHITIN

In many reports on chitin in fungi there are contradictory statements about its occurrence, probably because of application of detection methods of scarce reliability. Unlike other polysaccharides, chitin is insoluble in concentrated potassium hydroxide solutions at high temperature so that these conditions are realized in view of its deacetylation to yield chitosan, and generally the latter polymer is used for chitin detection.

With iodine–potassium iodide, chitosan assumes a brown colour which becomes red-violet on acidification with sulphuric acid.[47, 48] This is the Van Wisselingh test and enables chitosan to be differentiated from other material remaining after the alkali treatment. Another test is based on the treatment of chitosan with sulphuric acid to form spherocrystals of chitosan sulfate[49] which give a characteristic stain with fuchsin and picric acid.[47] Of course, it would be more correct to extract chitosan in formic or acetic acid. The use of these tests and of other less specific tests has been discussed,[50, 51] and it seems that negative results are not conclusive evidence of chitin absence, while positive chitosan tests are more reliable for indication of chitin presence.

Other methods based on the periodic acid Schiff reaction, and on chelation of metal ions, have been reported: the chelation of metal ions, namely iron, was mentioned to justify an unexpected positive reaction of the Hale reagent with chitin. Also the reaction of chitin with alcian blue containing copper phthalocyanin was erroneously reported as positive[53] and these data were defined anomalous by Salthouse[54] who, as well as Pearse, stated that the periodic acid Schiff, Hale and alcian blue reactions are negative for chitin. It was concluded that in the samples of chitin giving positive reaction other substances were present, probably proteins.[52]

Degradative methods for the detection of chitin have been applied in certain cases,[39, 55] but, since liberation of 2-amino-2-deoxy-D-glucose is indicative of chitin, only providing that there is a total removal of other materials containing this sugar, it is important to carry out an isolation of the polysaccharides.[56–58] Tracey[47] has indicated enzymatic methods for the qualitative detection of chitin.

X-ray diffraction[59–61] and electron microscopy[60] are the physical methods which are being used currently for reliable results, even though they require a certain amount of manipulation in order to clean the sample and to obtain readable X-ray diffraction patterns. The X-ray diffraction technique has been applied in research on chitins from various plant and animal samples, and for defining the crystalline forms of the polysaccharide.

Other instrumental methods include potentiometric determination of chitin: the method is based on the decomposition of chitin with concentrated sulphuric acid and the oxidation of the resulting glucosamine with potassium dichromate in excess; the excess is then titrated potentiometrically with the Mohr's salt.[62] A method based on deacetylation in hydrochloric acid was worked out for the quantitative determination of the content of N-acetylglucosamine and chitin in cell walls of yeasts. Acetic acid liberated in this reaction was determined by gas–liquid chromatography on a Porapak Q column.[63] The chitin chemistry and analysis were discussed in a review.[64]

ISOLATION OF CHITIN

Harsh treatments are required to free chitin from the accompanying materials.[65]

Chitin is usually prepared from crustacean shells by treatment with strong acids. Several procedures are listed. Rigby[66] treated crustacean wastes with hot 1% sodium carbonate solution followed by dilute hydrochloric acid (1–5%) at room temperature, and then 0·4% sodium carbonate solution. Blumberg[67] prepared chitin from lobster shells with hot 5% sodium hydroxide solution, cold sodium hypochlorite solution and warm 5% hydrochloric acid. Stacey[68] used 2 N hydrochloric acid at room temperature, and hot 1 N sodium hydroxide. The same was done by Hackman. The methods by Hackman,[69] Whistler and BeMiller,[70] Roseman *et al.*,[71] Foster and Hackman,[72] Takeda and Katsuura,[73, 74] and Broussignac[75] are here reported in detail.

Method of Hackman[69]

Lobster shells are cleaned by washing and scraping under running water, and dried in an oven at 100 °C. 220 g of the cleaned and dried shells are digested for 5 hr with 2 l of 2 N hydrochloric acid at room temperature, washed, dried and ground to a fine powder. The finely ground material (91 g) is extracted for 2 days with 500 ml of 2 N hydrochloric acid at 0 °C; the content of the flask being vigorously agitated from time to time. The collected material is washed and extracted for 12 hr with 500 ml of 1 N sodium hydroxide at 100 °C under occasional stirring. This procedure involving alkali treatment is repeated four more times. After washing to neutralize the reaction, chitin is washed with ethanol and ether and dried. The yield is 37·4 g corresponding to 17%, there are no ashes, and the nitrogen is 6·8% (calculated for $(C_8H_{13}NO_5)_n$ N = 6·89%).

Method of Whistler and BeMiller[70]

Lobster shells are cleaned by washing and dried in a vacuum oven at 50 °C. Five hundred grams of the clean dried shells are ground and soaked for 3 days in 10% sodium hydroxide solution previously deaerated, at room temperature. Fresh hydroxide solution is used each day. The deproteinized chitin is then washed until free of alkali, washed with 95% ethanol, 6 l being necessary to clean the product from pigments. Further, the chitin is washed with 1 l acetone, 2·5 l absolute ethanol, 500 ml ether. The white product is then dried under reduced pressure and introduced into a 37% hydrochloric acid solution at −20 °C for 4 hr. The particles are then washed with cold water. Washing with ethanol and ether conclude the procedure. The swelling in cold hydrochloric acid and following washings may be repeated. The yield is 100 g (20%), sulphated ash 0·15%, nitrogen 7·1%.

Method of Roseman and Coworkers[71]

Ten grams of decalcified lobster shells, prepared as described in the first part of the method of Hackman, are shaken for 18 hr with 100 ml of concentrated formic acid (90%) at room temperature. After filtration the residue is washed with water and treated for 2·5 hr with 500 ml of 10% sodium hydroxide solution on a steam bath. After filtration, the chitin is washed with water to neutral reaction, absolute ethanol and ether and dried under reduced pressure. The yield is 60–70% on the basis of decalcified shells, nitrogen 6·95%.

Method of Foster and Hackman[72]

This method is given here in detail because it seems important for understanding the chemical form of chitin in biological samples. In fact, the product obtained following the methods described above is termed chitin, but the drastic treatments, especially the prolonged extraction in hydroxide solutions often at high temperature, surely affect the chitin structure. It removes protein and peptides as desired, but it also removes acetyl groups and may well lead to fragmentation. The method by Foster and Hackman is a milder method for the isolation of chitin, based on the use of ethylenediaminetetraacetic acid on the cuticle of the edible crab *Cancer pagurus*. Selected, large cuticle fragments were attacked slowly (2 or 3 weeks) by EDTA at pH = 9·0. Powdered shells having particle size of 1–10 μm were decalcified more rapidly, in 15 min under the same conditions. No carbohydrate material was extracted in this process. After a further treatment with EDTA at pH = 3·0, the residue was extracted with ethanol for pigment removal and with ether for lipoid removal. Spectrographic results are as follows: (in parenthesis the values before treatment): SiO_2 1·07 (5·6); MgO 0·08 (4·5); CaO (77·0); P_2O_5 (8·20) % with an error of ±10–20%. Elemental analysis gave the following data, corrected for 2·6% ash as oxide: C 43·5; H 6·6; N 7·6. As indicated by the high nitrogen per cent, this chitin contained protein and this was demonstrated by hydrolysis in 6 N hydrochloric acid for 24 hr, followed by paper electrophoretic chromatography. Protein was roughly estimated to be 5% of the product. This protein could not be removed at room temperature by extraction with dimethylformamide or phenol–water mixture. Extraction with 98–100% formic acid gave a soluble and an insoluble fraction, both containing the same proportion of protein. The soluble fraction could be reprecipitated by dilution with water, and looks like chitosan. The insoluble chitin could be dispersed in aqueous lithium thiocyanate, saturated at room temperature at 100 °C; its fractionation gave fractions with similar protein contents. The protein content of chitin could be substantially reduced with repeated extractions with hot alkali.

The four procedures given above deserve some comments. First of all, if chitosan can be isolated after treatment with EDTA which should not produce any deacetylation, this would be in agreement with the statements of several authors that originally chitin is partially deacetylated, and a fraction rich in glucosamine residues can be separated by treatment with formic acid and reprecipitation upon dilution. For this main reason chitin cannot be produced pure, where the adjective pure refers to the prepoly-N-acetylglucosamine stoichiometry; moreover the treatments presented in the first three methods include prolonged digestion in hydroxide solutions, which are known to deacetylate chitin, and not only to eliminate protein. Therefore, in all cases a partially deacetylated product is obtained. This data is also of interest in connection with the production of chitosan: the ability of hydroxide solutions in eliminating proteins should allow one to care less for a preliminary protein elimination step, because the deacetylation process is based on hydroxide solutions at 120 °C, and therefore remaining aliquots of proteins are taken away at that stage. It is also remarkable the use of EDTA in both acidic and alkaline media: this treatment does not apparently deeply modify chitin as the reagent acts on the salts to which chitin is admixed, and thus justify the possibility of recycling of chitin and chitosan columns treated with EDTA.

Therefore, in the author's opinion, the treatment of cuticles should be different depending on the product desired—chitin or chitosan; in any case one should select clean cuticles from those parts of the body where no meat is attached to the cuticle, or cuticles from moulting.

Method of Takeda and Katsuura[73, 74]

King crab shells are decalcified with EDTA at pH = 10 at room temperature, and then digested with a proteolytic enzyme such as tuna proteinase at pH = 8·6 and 37·5 °C, papain at pH = 5·5–6·0 at 37·5 °C or a bacterial proteinase at pH = 7·0 at 60 °C for over 60 hr. The protein still present in chitin was about 5%. To remove the remaining protein, sodium dodecylbenzensulphonate or dimethylformamide were particularly effective.

Method of Broussignac[75]

The method of Broussignac is at present the best method available in the light of the knowledge reached about the raw material chemistry and performances of the finished products.

First of all, the crab shells are ground and sieved in order to obtain an homogeneous material, and to establish reproducible procedures and working schedules. Particles of a diameter of between 1 and 6 mm are suggested. On this material decalcification can be carried out by a simple treatment with hydrochloric acid containing about 50 g HCl l^{-1} at room temperature. This can be done in any container of plastic or wood; metals should be carefully avoided. Shaking is suggested, but when treating large amounts of crab shell powder, it is convenient to line up a series of containers and to pump the acid solution from the most decalcified chitin container to the least decalcified, in order to use the acid solution as completely as possible. It is not necessary to cool the containers. During decalcification the concentration of hydrochloric acid in each vessel is surveyed by acidimetric titration for a rational conduction of this operation. The titration is not disturbed by the presence of the small amount of phosphate in the shells. This operation takes about 24 hr: one should pay attention to the important gas evolution at the beginning which stops after 1 day. Before ending it is suggested to check the ash content which should be in the range of 0·4–0·5%. After completion of the decalcification treatment proteins are to be eliminated, and for production of chitin deacetylated as little as possible, papain, pepsin or trypsin are to be used.

When chitosan is the final product, the enzymatic treatment can be conveniently replaced by an alkali treatment at high temperature. At this stage a differentiation should in fact be made, depending on the desired final product. When chitin is not the final product, but an intermediate product in the chitosan production process, the alkali treatment for elimination of protein can be performed in a glass or steel container, steam heated at 90 °C. Five per cent sodium hydroxide solutions are used for a three-fold digestion of about 40 min. A water washing to neutral reaction yields chitin for chitosan production.

Pink or yellow pigments are often present in chitin, and sometimes one would appreciate having a chitin as white as possible. Pigments are destroyed by an oxidative treatment under very mild conditions. Chitin is soaked in a cold bath of acidic hydrogen peroxide for 6 hr. Four kilograms of hydrogen peroxide at 130 vol. with 10 kg hydrochloric acid 20 Bé are used to make up 200 l solution; which is enough for 10 kg chitin. Certain surfactants or soaps can also do this task. The total yield is about 10%.

Too little is known about formamide treatment at 130 °C followed by acetic acid 10% at 30 °C; this is reported to give chitin identical with that obtained by acid–alkali procedures.[76]

Chitin can also be isolated from the shell of the cuttlefish *Sepia officinalis*.[77] The shell of the cuttlefish is of elongated ovoid form with a hood-like structure at the anterior end. Sometimes commercially available

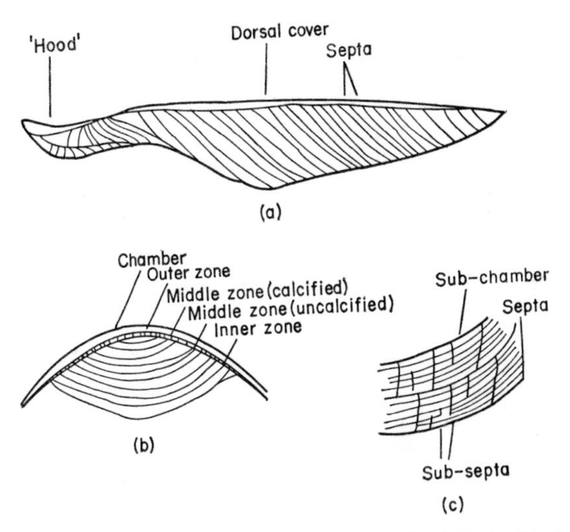

FIG. 3.1. Diagrams illustrating the anatomy of the shell of *Sepia*. (a) Diagrammatic longitudinal section of the shell of *Sepia*. In the adult shell about 100 septa are present; in this diagram only 28 are shown. (b) Transverse section of the shell of *Sepia*, showing the outer, inner, and middle zones of the dorsal cover; the middle zone is divided into a calcified and an uncalcified region (after Appellof). (c) Details of septa region showing the septa, sub-septa and the papillae which subtend the septa (after Denton and Gilpin-Brown). (From N. Okafor, *Biochim. Biophys. Acta* **101**, 193 (1965).)

shells lack this structure probably because of transportation. Anatomically, the greater part of the bone consists of chambers and subchambers bounded by septa, as shown in Fig. 3.1. The septa are overlain by a thick dorsal cover divided into three zones. The matrix of the septa and dorsal cover consist of chitin–protein complex. To isolate chitin the shell of the cuttlefish was first decalcified and the resulting chitin–protein complex was then deproteinized.

Dehooded shells were decalcified by treatment in 10% hydrochloric acid that was changed every day. The dorsal cover of the shell was peeled off on the 3rd or 4th day, and its inner layer removed by rubbing it away with fingers. The rest of the bone was broken up into smaller bits. Decalcification was complete in about 8 days. The resulting chitin–protein complex of the dorsal cover was deproteinized by treatment with 10% sodium hydroxide solution at 104 °C in an autoclave; the septa region was deproteinized under the same conditions in 6 hr. Ether extraction was done at the end of this operation. The dorsal cover chitin air-dried on glass plates was leathery and glass-clear where still in contact with the glass plate, otherwise translucent. Elemental analysis gave for dorsal cover chitin: C 46·3; H 6·3; N 6·73; ash 0·17. For septa chitin: C 46·7; H 6·2; N 6·61; ash 0·62. X-ray diffraction photographs of chitin from the dorsal cover and septa regions were identical. The *d*-values agree closely with those published for preparations of crab chitin and fungal chitin, except that while the innermost values for the two published accounts quoted were about 9·5 Å, in this experiment they were 11·0 Å. A dimension of about 11 Å is characteristic of β-chitin.

This technique and the results are of exceptional interest, as they provide a way of preparation of large chitin membranes which are not commonly available with this remarkable crystallinity degree. With some modifications this technique has been adapted for production of chitosan membranes that keep some degree of crystallinity. (See page 171).

Fungal chitins often have a nitrogen content lower than the theoretical 6·89%, and further purification was attempted using permanganate or bisulphite, which probably cause extensive degradation, and shortening of chain length. The same happens with purification by abrupt reprecipitation in concentrated mineral acids. The so-called colloidal chitins obtained by this procedure are sometimes used as enzyme substrates. A stable, colloidal dispersion of rod-like chitin crystallites, has been obtained from purified chitin heated under reflux with hydrochloric acid.

BIOSYNTHESIS

The activity of chitin synthetase, the enzyme which catalyses the synthesis of chitin from uridine diphosphate acetylglucosamine, has been studied in various stages of development of the aquatic fungus *Blastocladiella Emersonii*, in various fractions of both spores (which lack cell walls), and growing cells (which have cell walls).[78] Studies on bound glucosamine in this fungus, and on chitin biosynthesis on this substrate were also reported.[79-84] Enzymatic activity was found in preparations from crustacea and insects. The research by Camargo[78] on *Blastocladiella Emersonii* was concerned with variations in enzymatic activity during morphogenesis. When the enzyme obtained from growing cells was incubated, formation of a chromatographically immobile product was observed. The addition of acetylglucosamine, produced a 10-to 40-fold stimulation of the enzyme activity, as presented in Table 3.9, and Figs. 3.2 and 3.3. The amount incorporated reached a plateau in 30 min, as from

TABLE 3.9. EFFECT OF ACETYLGLUCOSAMINE AND OTHER SUBSTANCES ON CHITIN SYNTHESIS

Data are expressed as counts per min incorporated into product from UDP-^{14}C-GlcNAc in 30 min. All substances indicated were added to a final concentration of 0·02 M. The enzyme fraction employed was the 10,000×*g* pellet from growing cells.

Additions	Amount incorporated cpm
None	2,080
GlcNAc	16,690
Diacetylchitobiose	10,890
GlcNAc + diacetylchitobiose	21,670
Glucose	2,080
GlcNAc-1-P	2,438
N-Acetylmuramic acid	1,080
N-Laurylglucosamine	1,040
N-Acetylgalactosamine	2,280

(From E. Plessmann Camargo *et al.*, *J. Biol. Chem.* **242**, 3121 (1967).)

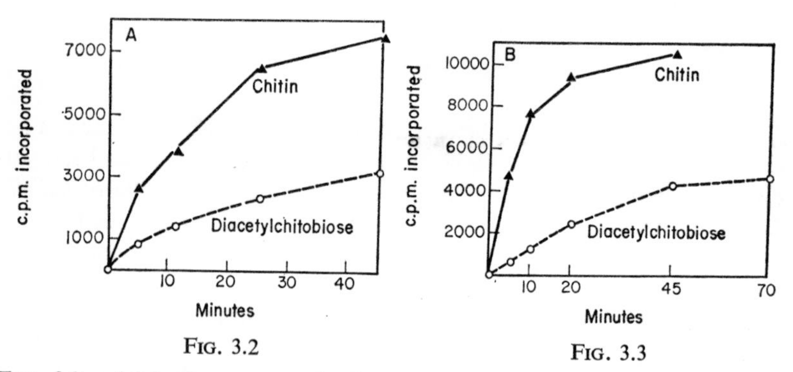

FIG. 3.2 FIG. 3.3

FIGS. 3.2 and 3.3. Time course of utilization of UDP-^{14}C-GlcNAc for synthesis of chitin and of diacetylchitobiose. Enzyme obtained after the disintegration of spores (FIG. 3.2) or of growing cells (FIG. 3.3) was employed. (From E. Plessmann Camargo, *et al.*, *J. Biol. Chem.* **242**, 3121 (1967).)

Fig. 3.4. The product of the reaction was identified as a chitin-like material by its susceptibility to hydrolysis by the enzyme chitinase. Both the polymer formed in the presence of UDP-^{14}C-NacGlc with unlabelled acetylglucosamine and that from unlabelled UDP-NacGlc with ^{14}C-acetylglucosamine were degraded by chitinase to yield two products, one with mobility of ^{14}C-acetylglucosamine and the other with the mobility of diacetylchitobiose that was also identified by chromatography in various

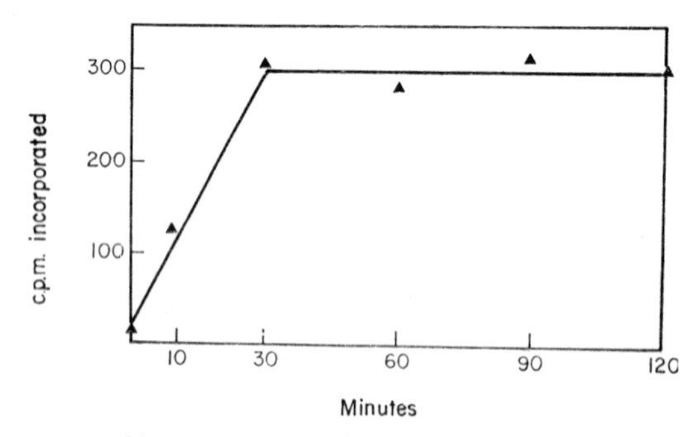

FIG. 3.4. Time course of incorporation of ^{14}C-GlcNAc into chitin the presence of unlabelled UDP-GlcNAc. The experiment was carried out as described in Table 3.9 employing ^{14}C-GlcNAc 20,000 cpm) and unlabelled UDP-GlcNAc at final concentrations 1×10^{-3} M and $2{\cdot}5 \times 10^{-3}$ M, respectively. (From E. Plessmann Camargo *et al.*, *J. Biol. Chem.* **242**, 3121 (1967).)

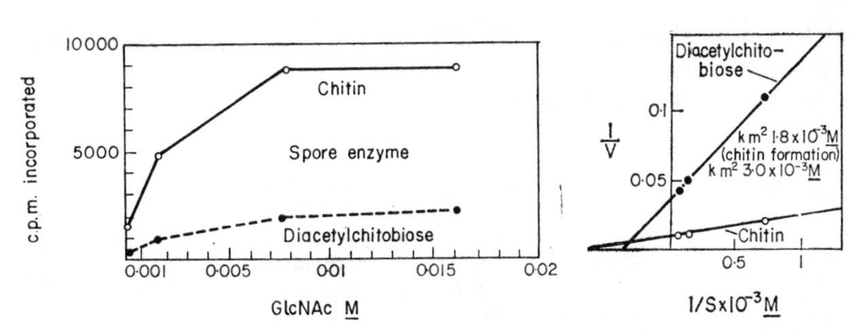

FIG. 3.5. Dependence of reaction velocity on the concentration of acetylglucosamine with enzyme obtained from spores. The substrate was UDP-^{14}C-GlcNAc.

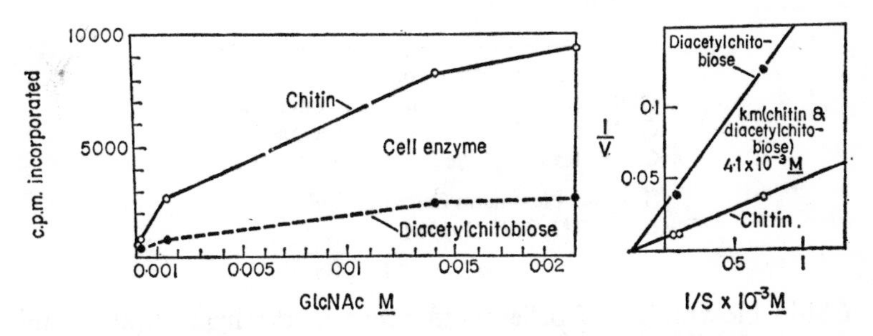

FIG. 3.6. Dependence of reaction velocity on the concentration of acetylglucosamine with enzyme obtained from growing cells. The substrate was UDP-^{14}C-GlcNAc. (From E. Plessmann Camargo *et al.*, *J. Biol. Chem.* **242**, 3121 (1967).)

solvents. The reaction velocity dependence on acetylglucosamine concentration for chitin formation is presented in Figs. 3.5 and 3.6.

Both the enzymes from the cell and that from the spores were inactivated by Co^{++} and Cu^{++}, and were activated by Mg^{++}, Ca^{++} and Mn^{++} to the extent reported in Table 3.10. Ca^{++} did not stimulate the enzyme from spores. The mechanisms of the above chitin synthesis reaction are not known, and the molecular weight was not determined.

Leloir and Cardini,[2] and Glaser and Brown[85] showed that chitin can be synthesized from UDP-N-acetylglucosamine by an enzyme from *Neurospora crassa*. Soluble chitodextrins stimulate the reaction and are converted to chitin. Data are reported in Table 3.4. A uridine nucleotide from the shore crab *Carcinus maenas* isolated by Kent and Lund takes part in the metabolism of chitin.[86]

TABLE 3.10. EFFECTS OF IONS ON CHITIN SYNTHETASE IN GROWING CELLS

Additions	Amount incorporated (rel. counting rate)
None	142
Mg^{++}	1363
Ca^{++}	322
Co^{++}	85
Cu^{++}	84
Mn	724

Assays were carried out with UDP-^{14}C-GlcNAc in the presence of unlabelled GlcNAc at 0·02 M final concentration. Concentration of the ion was $1{\cdot}5\times10^{-2}$ M.

(From E. Plessmann Camargo *et al.*, *J. Biol. Chem.* **242,** 3121 (1967).)

Chitin biosynthesis is believed to occur in the hypodermis, a thin cellular layer underlying the cuticle itself. The discovery in crustacean hypodermis of a nucleotide apparently identical with the uridine 5-(2-acetamido-2-deoxy-D-glucosyl dihydrogen pyrophosphate) of Glaser and Brown indicates the latter compound's involvement in the biosynthesis of chitin.[87] A similar nucleotide was reported to be present in moth haemolimph[88] and in extracts of locust wings;[89] synthesis of chitin however, could not be induced by the wing extracts.

The cuticle of arthropods is continuously digested, reabsorbed and formed again at each moult, so the system of biosynthesis and degradation of chitin in arthropods is very active. Moulting is followed by an intense period of chitin synthesis at the expense of glycogen by way of simple D-glucose derivatives, and conversely in the pre-moult period, synthesis of glycogen was observed to occur at the expense of cuticular chitin.[89–94]

TABLE 3.11. ENZYMATIC SYNTHESIS OF CHITIN FROM RADIOACTIVE UDPAG

Expt.	^{14}C-labelled compound added to incubation mixture	Position of label	Incubation time (min)	Total ^{14}C in insoluble polysaccharide* (cts/min)
A	UDPAG**, 3·8 µmoles, 76,560 cts/min	Acetyl	0	40
	–same–	–same–	150	16,900
	UDPAG***, 1·4 µmoles, 40,185 cts/min	Acetyl	120	2,063
	UDPAG***, 1·9 µmoles, 50,490 cts/min	Sugar	120	2,870
B	AG-6-P***, 1·7 µmoles, 165,480 cts/min	Acetyl	120	6
	(AG-1-P+AG-6-P)***, 1·2 µmoles, 123,125 cts/min	Acetyl	120	20
C	UDPAG§, 1·3 µmoles, 27,040 cts/min	Acetyl	120	1,200
	AG§ 1·0 µmole, 20,000 cts/min	Acetyl	120	0

* The polysaccharide was isolated by making the reaction mixture 1 N in $HClO_4$. The precipitate was collected and washed four times with 2 ml portions of 0·3 N $HClO_4$ followed by one washing with 2 ml of water. For counting, the precipitate then was suspended in water and a suitable aliquot plated.

** Reaction mixture contained 115 mg soluble chitodextrins, 30 µmoles $MgCl_2$, 3 µmoles EDTA (versene), 150 µmoles Tris, pH 7·5. Enzyme from 4·2 g *Neurospora*. Final vol. 5·8 ml 25 °C.

*** Reaction mixture contained 2·5 mg chitodextrins, 5 mg glutathione, 50 µmoles $MgCl_2$, 15 µmoles EDTA, 250 µmoles Tris. Enzyme from 2·5 g *Neurospora*. Final vol. 5·8 ml 25 °C.

§ Reaction mixture contained 8 mg chitodextrins, 20 µmoles $MgCl_2$, 2 µmoles EDTA, 100 µmoles Tris. Enzyme from 1·5 g *Neurospora*. Final vol. 4·5 ml 25 °C. In the experiment with labelled AG, 1·5 µmoles of unlabelled UDPAG were present.

(From L. Glaser *et al.*, *Biochem. Biophys. Acta* **23,** 449 (1957).)

PROPERTIES OF CHITIN

Chitin is a white solid insoluble in water, dilute acids, cold alkalis of any concentration, and organic solvents. Chitin can be dissolved with difficulty in liquid ammonia; Schweizer's cuprammonia reagent does not dissolve the polysaccharide, presumably because the acetamido group at C-2 prevents formation of the complex.[46] Chitin is sometimes reported to dissolve in hydrochloric acid or sulphuric acid, but in practice it undergoes hydrolysis under limiting conditions as reported by Hackman, and therefore, under normal conditions in dilute acids chitin is not appreciably soluble in a short contact time: in fact for isolation of chitin from crab shells 2 N hydrochloric acid at 25 °C is used for 2 hr.[95, 96, 68]

To measure the hydrolysis extent with hydrochloric acid *Lucilia* and *Loligo* chitins were heated at 100 °C with 2 N hydrochloric acid in sealed tubes for 1·5 and 24 hr.[95] The hydrolysates were cooled and their refractive indices measured in a differential refractometer with 2 N HCl as the reference liquid, and compared with that of N-acetyl-D-glucosamine, which had been hydrolysed in the same manner. The method gave an estimate of the total amount of chitin hydrolysed and the results are given in Table 3.12. The refractive index of an 0·5% solution of glucosamine hydrochloride in 2 N HCl was not altered after it had been heated in a sealed tube at 100 °C for 24 hr.

TABLE 3.12. ACIDIC HYDROLYSIS OF CHITINS
Chitin (particle size 150–200 mesh, 50 mg) in 2 N HCl (10 ml) at 100 °C

Period of digestion (hr)	Chitin hydrolysed (%)		
	Lucilia cuprina	*Loligo australis*	*Scylla serrata*
1	13·8	32·9	15·9
5	18·5	49·9	
24	27·5	88·7	

(From R. H. Hackman *et al., Austr. J. Biol. Sci.* **18,** 935 (1965).)

Fifty per cent nitric acid, 85% phosphoric acid[97] and anhydrous formic acid were used to perform chitin dissolution; therefore, the term dissolution is inappropriate because hydrolysis, molecular weight degradation, deacetylation, and chemical modification (sulphation, nitration etc.) take place under the said conditions and yield dissolved species which are no longer chitin.

It is a well-known fact that boiling concentrated hydrochloric acid eventually hydrolyses the ether and amide linkages in chitin to give glucosamine, but many workers believed that in cold hydrochloric acid no hydrolysis occurs and have used this reaction to purify chitin. Solutions in acids were tried for obtaining threads of chitin but without satisfactory results. Sheets of chitin were obtained by reprecipitating it from hydrochloric acid solutions and allowing it to dry on glass. Its tensile strength was reportedly so great that it pulled bits of glass from the plate surface, but of course it was readily soluble in water. The purification of chitin by dissolution in hydrochloric acid and reprecipitation is of course unsatisfactory.[98]

The above mentioned dissolution process takes place when hydrochloric acid normality is about nine, while in sulphuric acid the normality must be considerably higher, and concentrated nitric acid has practically no effect. This information is important in order to plan chromatographic applications of chitin in acidic media.

By dissolving chitin in fuming nitric acid, and allowing it to remain for 1 or 2 hr, and reprecipitating by stirring into water, chitin nitrate can be prepared. There is some suspect that nitrogen can enter in forms different than nitrate, because of limited hydrolysis or oxidation of chitin. A typical amorphous pattern was recorded by X-ray diffraction. Chitin nitrate exhibits the same solubility in hydrochloric acid as the original chitin, so that it is unlikely that any hydrolysis of the acetyl groups has occurred. In hydrochloric acid solution there is some denitration, as revealed by X-ray diffraction. Nitration of a fibre of chitin was also attempted for structure studies. This was done with 5 parts of fuming acid and 1 part of concentrated nitric acid. The fibre was extracted with formic acid to remove short-chain fractions. Chitin nitrate was said to be orthorhombic with $a = 9$, $b = 10{\cdot}3$ and $c = 23$ A.U. where the c distance is increased as expected from the substitution of a hydroxyl by a nitrate group.[98]

Current recipes for preparation of D-glucosamine hydrochloride include treatments of dry chitin with concentrated hydrochloric acid (sp. gr. 1·18) heated on boiling water for 2·5 hr. Treatment of glucosamine hydrochloride with strong bases such as diethylamine or triethylamine gives free D-glucosamine.[99, 96]

The graded acidic hydrolysis of chitosan (de-N-acetylated chitin) yields a series of saccharides which may be fractionated by elution from ion-exchange resins.[68, 71] A partial degradation of chitin which permits the isolation of a series of N-acetylated chitosaccharides up to at least the heptasaccharide was also reported by Barker, and coworkers.[68] Ten grams of chitin were added to a cooled mixture of 50 ml acetic anhydride and 6·5 ml concentrated sulphuric acid. After storage for about 2 days at controlled temperatures the mixture was added to a cooled (0 °C) solution of sodium acetate trihydrate in water (40 g in 260 ml). The supernatant liquid was neutralized and extracted with chloroform. The extract was washed with water, dried, and the residue was recrystallized from methanol to yield 1·14 g of chitobiose octo-acetate as colourless needles melting at 286–288 °C.

Data were published on chromatography of chitin oligosaccharides using a mixture of pentan-2-ol-pyridine and water, followed by detection with alkaline silver nitrate. Isoamyl alcohol+pyridine+water (1:10·8) was claimed to give better results on paper. The paper was then exposed to HCl vapours and sprayed with starch-iodide reagent. Sensitivity was less than 1 μg.[100]

Acetolysis of chitin has been described by other authors too. The value of acetolysis for degrading chitin is limited by the formation of deacetylation artefacts, although the effect can be minimized by carrying out the deacetylation with magnesium methoxide.[55–58, 101, 102]

Jeanloz and Forchielli[97] studied the oxidation of chitin by periodate: they proposed the following mechanism:

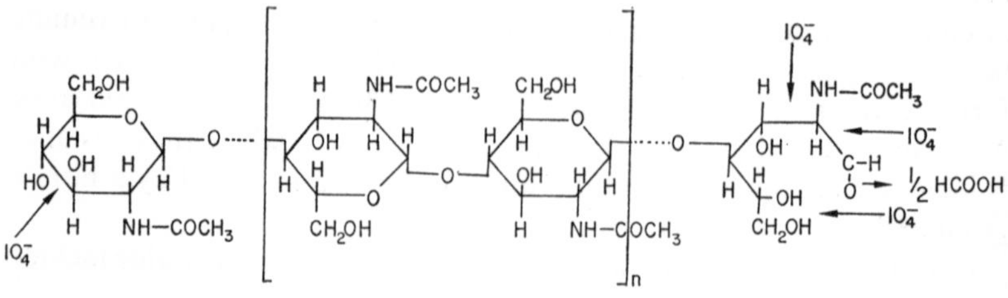

The product resulting from a prolonged treatment at 25 °C and pH = 7·3 was isolated. Nitrogen and acetyl groups contained in this product are higher than in chitin, respectively 7·4 and 24·0%. Under those conditions periodate attacked bonds different than C_1–C_2 and C_2–C_3, because, otherwise nitrogen per cent would become lower. The probable way of this reaction is a hydroxyl group introduction on C_5 followed by the C_5–C_6 bond rupture.

Chitin and several other high polymers can be dispersed in hot concentrated solutions of neutral salts capable of a high degree of hydration, and they may be reprecipitated in form of filaments by pouring these dispersions into alcohol.[98] Chitin can be dispersed in a syrupy colloidal solution by heating at 95 °C a lithium thiocyanate solution saturated at 60 °C. As a precipitation medium acetone in water was found to be superior to alcohol, since the speed of the precipitation could be varied between wide limits by merely changing the proportions of acetone. By extrusion of the solution through a tip into acetone and water in mixture, threads could be made. They developed a remarkable degree of orientation by tension application, as visible between crossed Nicols. Some of the jelly-like mass of chitin precipitated from the colloidal solution by dilution was allowed to dry on a glass plate. The diffraction pattern obtained with the beam parallel to the surface of the sheet showed the same type of orientation as the natural sheet. Chitin precipitated from several-months-old lithium thiocyanate solutions, showed no evidence of hydrolysis from its diffraction patterns. Lithium iodide can replace lithium thiocyanate, but its action is slower.

The effect of these salts is to force the chains apart without breaking them and without producing any displacement in the direction of the *b* axis, as the interferences of planes perpendicular to the chains, 030, 031 remain sharp. If chitin is allowed to remain in the salt solution long enough, the chains are spread so far apart that the polymer goes into solution.

Chitin nitrate can be dispersed in lithium thiocyanate and reprecipitated without alteration.

The dispersal efficiency follows the orders:[93, 103]

$$LiCNS > Ca(CNS)_2 > CaI_2 > CaBr_2 \quad \text{and}$$

$$LiCNS > LiI > LiBr > LiCl.$$

Dispersion of chitin in aqueous calcium chloride cannot be effected.

ENZYMATIC DEGRADATION

Bacteria which completely attack chitin are called chitinoclastic bacteria, to distinguish them from bacteria that affect only the extremities of the polymer chain. In surface sea-water less than 1% of the microorganisms have chitinolytic activity.[18, 104]

Chitin is attacked to yield ammonia, acetate, glucosamine and sugars. In the culture media, the ammonia evolution is rather quick at 21 °C.[105, 106]

The amino groups are first hydrolysed:

$$\equiv C—NHCOCH_3+H_2O \rightarrow \equiv C—OH+CH_3CONH_2$$

$$CH_3CONH_2+H_2O \rightarrow CH_3COOH+NH_3.$$

Ammonia is partially used as a nitrogen source, and acetic acid enters the metabolism immediately upon formation. In the open sea in temperate climate, chitinolytic activity is very limited. In the Mediterranean Sea, especially along the African coasts, the chitonoclastic bacteria are more abundant.

Bacteria attached to plankton are less effective than expected, due to the fact that living plankton carries different bacteria than dead plankton. Gram-negative bacteria, particularly vibrios, *Pseudomonas*, are by far the richest ones in chitinoclastic enzymes.[107–109] Chitinases can be obtained from the higher fungi and *Lycoperdon* sp. (puffballs) are a convenient source for large amounts of high activity enzymes.[110]

Chitin prepared from crab shells was added to a basal medium consisting of 0·1% glucose, 0·1% peptone, 0·005% dipotassium hydrogen phosphate and 1·5% agar in sea-water.[111] The cultures were streaked on the surface of this medium, incubated and examined for the presence of a zone of clearing of the chitin around the bacterial growth. It appeared that chitinoclastic activity is widespread among the luminous bacteria (photobacteria). The part that the luminous bacteria play in the chitin degradation is not clear. Certain luminous bacteria have been found in the bacterial flora of the intestines of certain fishes whose diet may rely considerably on crustacea. There is no information on the occurrence of photobacteria at great depths where chitin finds its way to the sea bottom.

Berger and Weiser[112] reported that crystalline, eggwhite lysozyme degraded chitin but was without action on chitosan.

Also partially *O*-carboxymethylated chitin was found to be a soluble substrate of egg white lysozyme.[113] The products of the lysozyme action were analyzed by ion-exchange chromatography and gel filtration.

MACROSTRUCTURE

Chitin is a crystalline polysaccharide, as demostrated by X-ray diffraction, infrared sprectrophotometry, isotope exchange, thermal analysis, hydrolysis and chromatography data. Research in this field has been helped, in recent years, by the isolation of fibres of sure composition, amongst which is remarkably, the so called "chitan" from *Thalassiosira fluviatilis*. Some essential information is reported below, in order to explain the substantial findings of many authors engaged in the study of the macrostructure of this polysaccharide, which appears to be the most examined after cellulose. The data available constitutes a rather complete body of knowledge, on which the technical and analytical applications of chitin and its derivatives can be based. In particular, the crystallinity parameters should be taken into account for the understanding of the interactions of this class of chelating polymers with metal ions, from the stereochemical point of view, as one can reasonably expect that a convenient steric disposition of complexing groups would enhance the bond formation capacity of the chelating polymer.

It is also interesting to remark the correspondence existing between the presently available information on macrostructure of synthetic and natural chelating polymers: in both cases the interest in the stereochemical characteristics has been developed, but while for synthetic polymers this was done by organic chemists in view of the synthesis of selective resins, the macrostructure of chitin and derivatives was mainly studied by physiologists, oceanographers, biophysicists and entomologists with no direct interest in the interactions of the polymers with metal ions.

Chitin is at present considered to exist in three distinct crystalline forms, the so called α, β and γ chitins.

Detailed structural work has been reported especially for α and β chitins.[114] Fundamental to these proposed structures has been that parallel chitin chains are arranged in bonded "piles" or "sheets" linked by N—H···O=C hydrogen bonds through the amide groups, as indicated in Fig. 3.7, and that the α and β forms differ in that the piles of

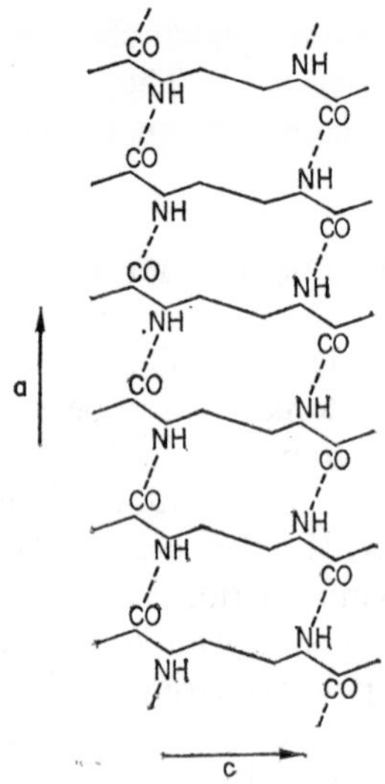

FIG. 3.7. *Pile* or *sheet* of chitin chains viewed along the fibre axis: (a) hydrogen-bond direction, (c) side-chain direction. (From J. Blackwell, *Biopolymers* **7**, 281 (1969).)

chains are alternately antiparallel in α chitin, and parallel in β chitin. Rudall has suggested that in the γ form the piles of chains are arranged in sets of three with two parallel chains and one antiparallel (figs. 3·8 and 3·9).

A structure for chitin was proposed by Dweltz[115] on the ground of X-ray diffraction studies carried out on lobster tendons; his conclusions were for a well-defined orthorhombic lattice, with the space groups $P2_12_12_1$ and cell dimensions of $a = 4{\cdot}69$, $b = 19{\cdot}13$ and $c = 10{\cdot}43$ Å (fibre repeat). He suggested that the unit cell contains two polysaccharide chains running in opposite directions and four asymmetric N-acetylglucosamine units which give a density close to the experimentally observed value of 1·42 g ml^{-1}. The structure was established by a trial and error method with the aid of an optical diffractometer for testing the diffraction patterns of the various possibilities suggested by stereochemical considerations. The chains are separated by a distance of 4·69 Å perpendicular to the "plane" of the sugar ring and consequently the NH and CO groups are hydrogen bonded, the plane of the amide group in the "trans" configuration being almost perpendicular to the fibre axis. The hydroxymethyl side chains are

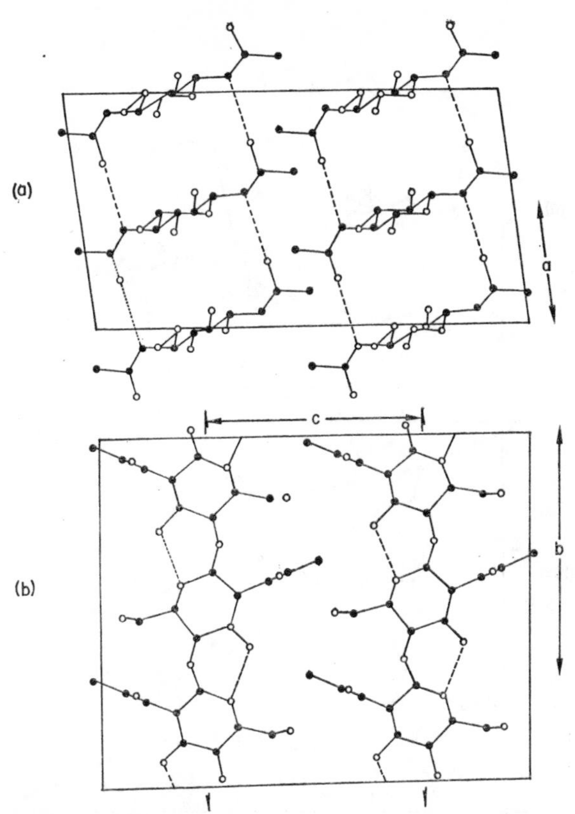

FIG. 3.8. The β-chitin structure (Blackwell): (a) projection on ac planes; (b) projection on bc plane. (From K. M. Rudall, *J. Polymer Sci.* **28**-C, 83 (1969).)

also hydrogen bonded, the OH of one being to the oxygen of a similar group from a neighbouring chain running in the opposite direction.

In earlier studies Meyer and Pankow[116] too deduced that the cell is orthorhombic, but with parameters $a = 9{\cdot}40$, $b = 10{\cdot}46$ (fibre repeat), and $c = 19{\cdot}25$ Å, with $P2_122_1$ space group. Other works confirmed these findings. Postulating an analogy between cellulose and chitin, Meyer, Mark, and Wehrli[117, 118] arrived at the spatial formula of chitin where the pyranose rings are linked by a $1 \rightarrow 4$ type of bonding. Meyer and Pankow introduced the concept of the chitobiose unit, analogous to the cellobiose unit in cellulose, in which they postulated that the two adjacent

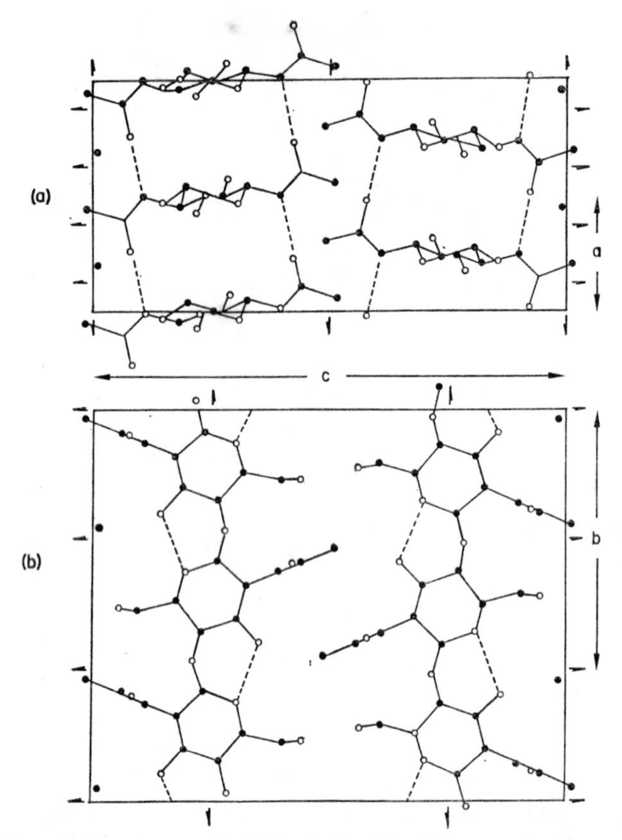

FIG. 3.9. The *a*-chitin structure (Carlström): (a) projection on ac plane; (b) projection on bc plane. (More than the unit cell is shown in Figs. 3.8 and 3.9; the lengths of the cell axes are given, a, b, c.) (From K. M. Rudall, *J. Polymer Sci.* **28-C**, 83 (1969).)

pyranose rings along a chain were similar but not crystallographically identical, and therefore in the unit cell occurred four chitobiose units.

Using chitin from the apodemes of the lobster *Homarus americanus* Carlstrom[119] obtained an orthorhombic unit cell having dimensions of $a = 4{\cdot}76$ Å, $b = 18{\cdot}85$ Å and $c = 10{\cdot}28$ Å (fibre repeat); this structure is quite similar to that of cellulose. Carlstrom did not accept a straight arrangement of the pyranose rings in the disaccharide units, and favoured a bent-chain structure whose features include intramolecular hydrogen bonding between the C-3 hydroxyl groups, and the ring oxygen of the next sugar residue. This structure was supported by polarized infrared measurements on various chitin samples where neither free OH or NH or C=O···HO bonding was found.[120, 121]

In discussing the Dweltz and Carlstrom structures it was concluded that from X-ray diffraction recordings no decision can be taken about a choice of one of the two structures.[93]

Many doubts on both the above structures were cast by Falck and coworkers[122] who criticized the preparation methods for the lobster fibres which generally included acid and alkali treatment. Extensive changes in the polymer structure can be brought about by these treatments, and even boiling in pure water produces extended changes. The macrostructure of chitin as previously isolated may well be an artifact formed in the process of isolation, and therefore comporting difficulties in reproduction and comparisons. The three types of chitin may also be unrelated to the real state of chitin in the organisms.

Therefore those Canadian authors[123] isolated and described fibres composed of pure crystalline poly-N-acetylglucosamine from *Thalassiosira fluviatilis*. No other information about occurrence of chitin in diatoms has been made, although certain authors considered this point. This substance was called chitan. The name chitin was reserved for the polyaminosugar obtained from other sources, and accompanied by other materials, so that it would not be pure and completely acetylated and quite crystalline.

An axenic culture of the euryaline centric diatom *Thalassiosira fluviatilis* was used. The medium developed an apparently high viscosity, even at relatively low cell density. The increased viscosity resulted from an abundant production of numerous long fibres attached to the valve surfaces of the cell. Synthesis of chitan, the constituent of the frustules, was relatively rapid. The fibres were only superficially associated with the protoplast. They were 60–80 μ long whereas the thickness was less than 1 μ; they were quite rigid. From X-ray data it was calculated that the number of repeating units of N-acetylglucosamine in a chitin microfibril was nearly 1000 μ^{-1}, therefore, assuming a single polymer chain, these microfibrils should have contained $6\text{–}8 \times 10^4$ repeating units.

Hydrolysates of the fibres revealed only glucosamine. The authors reported that this was a real homoglycan, because chitin samples previously considered were not a distinct chemical entity, but always a component of a chitin–protein complex. In both fungi and algae, chitin has been considered as a component of the cell wall. Chitan from *Thalassiosira fluviatilis* was not found in the cells and it was concluded that chitan was restricted to the fibres and was not part of the cell wall.

Results of chemical, proton magnetic resonance (p.m.r.), infrared, and X-ray diffraction studies were published to establish the composition and structure of chitan, and provide information on its macrostructure.[122] The difference between chitan and arthropod chitin was also taken into account.

In contrast to the chemical or enzymatic methods of preparation of chitin, the purification of chitan consisted only of a solvent wash and drying *in vacuo*. The elemental analysis of these fibres conformed to $(C_8H_{13}O_5N)_n$.

Proton magnetic resonance

Proton magnetic resonance studies on the acid hydrolysis of chitan in deuterium oxide to suppress signal disturbance enabled the number of hydrogen atoms directly attached to the carbon of each type in the repeating unit to be determined, and to understand the hydrolysis mechanism.

The signals in the spectrum of chitan, reported in Fig. 3.10, could be assigned to the anomeric hydrogen on C-1 (4·80) of each repeating unit, the four hydrogens in the pyranose ring together with those of the two methylene hydrogens at C-6 (5·70), and the three methyl hydrogens of the N-acetyl group at C-2 (7·10). Slight hydrolysis of chitan has also given rise to smaller signals for the methyl hydrogens of acetic acid (7·47) and the α-anomeric hydrogens of N-acetylglucosamine (4·25). The signal for the β-anomeric hydrogen of the latter was hidden under the envelope at 4·80, and that for the anomeric hydrogens of glucosamine was too weak to be observed. Figure 3.10 (b) effectively illustrates the conversion of chitan into glucosamine and *N*-acetylglucosamine. The signals for the α-anomeric hydrogens of *N*-acetylglucosamine and glucosamine were doublets centred at τ 4·25 and τ 4·04, whereas the corresponding doublets for the β-anomers occurred at higher fields and were centred at τ 4·65 and 4·49. These assignments were based on the spectra of the pure compounds and the hydrolysis products of an authentic sample of N-acetylglucosamine in Fig. 3·10 (c). The pyranose ring and methylene hydrogens on C-6 of glucosamine and its *N*-acetyl derivative appeared at a higher field with an average τ value of 5·72, and

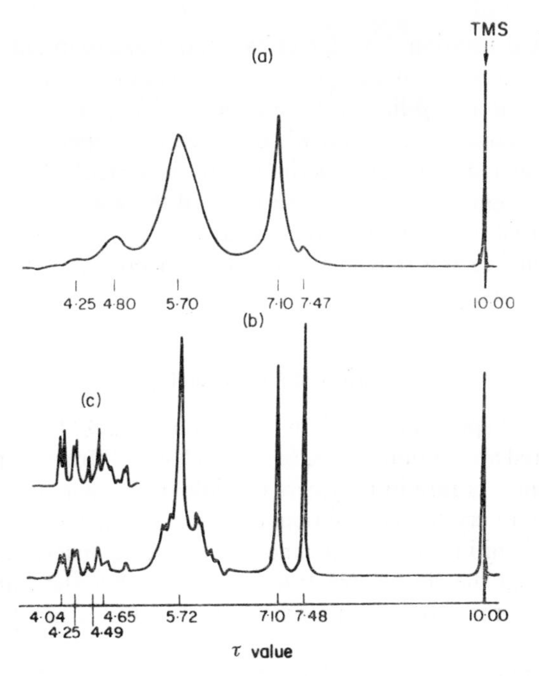

FIG. 3.10. The p.m.r. spectra of chitan after 1·7 hr (a) and 32 hr (b) in 10·2 N HCl at 40°. A section of the spectrum from the hydrolysis products (*ca.* 4 hr at 50°) of an authentic sample of *N*-acetylglucosamine is shown in the insert (c). TMS = tetramethylsilane. (From M. Falk *et al.*, *Can. J. Chem.* **44**, 2269 (1966).)

the methyl hydrogens of the *N*-acetyl group and acetic acid could be assigned to the sharp singlets at τ 7·10 and 7·48. No signals were obtained for any hydroxyl protons because of exchange with deuterium oxide. It should be noted that the p.m.r. evidence established that *N*-acetylglucosamine, glucosamine, and acetic acid were the only products of hydrolysis, and the use of an internal standard established that one mole of acetic acid was produced per $C_8H_{13}O_5N$ repeating unit. Glucosamine hydrochloride could also be isolated in an excellent yield after the hydrolysis reaction was complete. Results obtained for the hydrolysis of crustacean chitin under the same experimental conditions were identical with those reported above, except that the yield of acetic acid was significantly lower than the theoretical amount.

The optical rotation data for chitan and crustacean chitin were also indicative of a β-D-configuration in these compounds, and it was established that chitan is β-linked 2-acetamido-2-deoxy-D-glucan.

Chitan remained unchanged when heated at temperatures of 300 °C. It was calculated but not experimentally measured that chitan had a molecular weight of 3×10^7 on the basis of the average length of 80 μ for the microfibres, and an estimated value of 10 Å for the length of the repeating unit in the polymer chain. Chitan does not dissolve in any organic solvents.

Infrared spectrometry

The infrared spectrum of chitan is shown in Fig. 3.11. Chitan has very sharp infrared absorption bands, as well as X-ray diffraction patterns.[122] Such sharpness is rare in polymers of biological origin and it is characteristic of pure crystalline compounds. As it is known the more diffuse spectrum of chitin indicates a much less ordered macrostructure. One would say that the chitan structure is perfect, while the chitin structure has been disturbed to a certain extent. In fact the infrared spectrum of

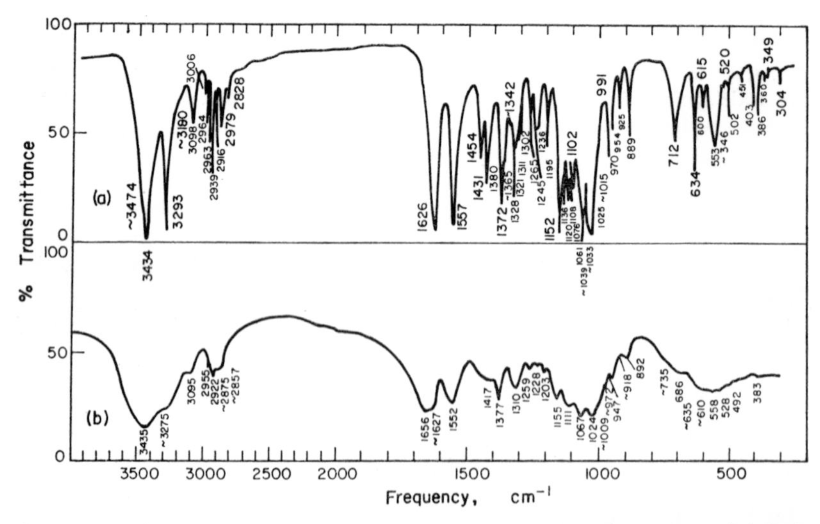

FIG. 3.11. The infrared absorption spectrum of chitan (a) and chitin (b). (From M. Falk *et al.*, *Can. J. Chem.* **44**, 2269 (1966).)

chitin is representative of the macrostructure of the fully *N*-acetylated regions of the substance, which, as seen before, is not homogeneous because it was estimated that a certain degree of deacetylation occurred.

The conclusion that chitan, but not chitin, is completely crystalline is supported by hydrogen–deuterium exchange experiments. When mats of chitan were immersed in deuterium oxide at 21 °C, only about 25% of the hydroxyl protons and 6% of those in the NH groups were exchanged after 1 month. The lithium thiocyanate treated mats exchanged many times faster. In fact by treating chitan with thiocyanate, an infrared spectrum similar to that of chitin could be obtained, so demonstrating that chitan can be transformed in chitin by a method that favours deacetylation. Analogous experiments with cellulose, have established that the rate of deuterium exchange is much faster for amorphous than for crystalline cellulose.

The infrared spectrum of chitan in Fig. 3.11 shows two absorption bands at 3474 and 3434 cm^{-1} caused by the OH stretching vibrations of the two hydroxyl groups in each *N*-acetylglucosamine residue. Deuterium exchange caused a slow diminution of both bands at about the same rate, with the concomitant appearance of the bands of the two corresponding OD vibrations at 2565 and 2545 cm^{-1}. The absence of absorption above 3500 cm^{-1} in chitan shows that there are no "free" hydroxyl groups, i.e. that all of the hydroxyl groups in chitan are involved in hydrogen bonds.

The sharpness of the OH stretching bands is particularly remarkable. The half-widths of the OH bands (about 20 cm^{-1}) and of the corresponding OD bands (about 10 cm^{-1}) are smaller by an order of magnitude than those of OH bands in the spectra of most biopolymers. Indeed, these may be the sharpest such bands ever observed for hydrogen-bonded OH groups at room temperature. The sharpness of the OH absorption bands shows that each type of hydroxyl group in chitan, in contrast to chitin, has an identical environment.

Amide groups in chitan give sharp infrared bands at 3293, 1557, and 712 cm^{-1} which are due to the NH stretching vibration, the amide II vibration (largely the in plane NH bending), and the amide V vibration (largely the out-of-plane NH bending). The behaviour of chitan is similar to that of other hydrogen-bonded secondary amides such as proteins and polypeptides[124] which exhibit slow rates of isotope exchange of the NH protons in the presence of deuterium oxide at room temperature. For example, only 6 per cent of these protons in chitan are replaced by deuterium after 1 month at 21°, although there is a marked temperature dependence which results in the exchange of almost 40 per cent of the NH protons in 2 hr at 100°. The diminution of the NH bands is accompanied by the appearance of the corresponding NH vibrations at 2462–2428 (doublet), 1479, and 522 cm^{-1}. Doublet formation in the ND stretching region is probably due to Fermi resonance. These frequencies are characteristic of hydrogen-bonded NH groups in polyamides and polypeptides[125, 126] and clearly establish the absence of free NH groups. Frequencies similar to those discussed above are observed in the spectrum of chitin, although the bands are much broader.

One of the most distinctive features of the infrared spectrum of chitan is the single, sharp, intense band at 1626 cm^{-1}, which is due to the amide I vibration (largely C═O stretching). The sharpness of the band indicates immediately that the amide groups in this polysaccharide are in an equivalent environment. Crystalline *N*-acetylglucosamine also exhibits a single absorption band at 1626 cm^{-1}, and so does its 2-epimer *N*-acetylmannosamine at 1623 cm^{-1}. It seems probable, therefore, that this band position is characteristic of monomeric or polymeric 2-amino-2-deoxyglycopyranoside units in the crystalline state. The position of the amide I band in this class of compounds appears to be insensitive to the configuration at C-2 of the hexosamine residue, and suggests that the factors influencing this mode of vibration are largely confined to the amide group itself. An X-ray study[127] indicates that the plane of the amide group in crystalline *N*-acetylglucosamine is perpendicular to the plane of the pyranose ring of the hexose residue, and that a network of hydrogen bonds exists between the NH and C═O groups in neighbouring molecules. An analogous interchain system of hydrogen bonds is likely to be present in chitan. Certainly, the characteristic vibrations associated with the amide groups in the latter are consistent with the presence of a hydrogen-bonded secondary amide,[125, 126, 128] and the presence of a system of C═O···HN hydrogen bonds in a plane perpendicular to the plane of the pyranose ring is supported by a consideration of the molecular models.

In contrast to the spectrum of chitan, which has only one band in the C═O stretching (amide I) region, that of chitin contains bands at 1626 and 1656 cm^{-1} which have been the source of some disagreement. The band at higher frequency has previously[119–121] been attributed to the amide I vibration. On the other hand, the band at 1626 cm^{-1} has been assigned to a C=N stretching vibration,[121] an overtone or combination mode,[122] an amide C=O group hydrogen-bonded to an OH group[129] or the C=O group of a rotational isomer of the aminoacetyl group,[121] or has been completely ignored.[115] The present evidence strongly indicates that these two bands are in fact both due to amide I vibrations, but correspond to two different classes of amide groups. This conclusion is supported by the following observations. (a) Chitan has only one band, that at 1626 cm^{-1}, and crystalline *N*-acetylglucosamine and *N*-acetylmannosamine also have a single amide I band at very nearly the same frequency. (b) Treatment of chitan with lithium thiocyanate and with other reagents causes a simultaneous appearance of the new band at 1656 cm^{-1}, and a decrease in the intensity at 1626 cm^{-1}. (c) Reprecipitation of chitin from hydrochloric acid causes an intensification of the band at 1656 cm^{-1}, with a corresponding decrease at 1626 cm^{-1}. (d) The band at 1626 cm^{-1} has been observed to disappear more rapidly than does the band at 1656 cm^{-1} on progressive deacetylation and during nitration.[129] (e) The two bands differ in their relative intensities and dichroism in different preparations of chitin.[119–121, 129, 130]

It has been suggested that the band at 1656 cm^{-1} in chitin may be due to water of hydration.[131] However, it was found that prolonged drying at 110° under vacuum has no effect whatsoever on the intensity of this band in samples of crustacean chitin or of lithium thiocyanate treated chitan. The reported diminution of this band upon partial deuteration of chitin[131] was confirmed[122] but may correspond to the usual decrease in the frequency of an amide I band upon deuteration.[126, 128] There was certainly no appearance of the corresponding bands of HDO near 1450 cm^{-1} and of D_2O near 1200 cm^{-1} as deuteration proceeded. Thus the band at 1656 cm^{-1} cannot be due to water of hydration.

The amide I vibrations of neighbouring amide groups within a unit cell are often strongly coupled, in which case the coupled vibrations may be removed as much as 20 cm^{-1} from the frequency of the uncoupled vibrations.[132–134] It is thus possible that the band at 1626 cm^{-1} in chitan (as well as the band at 1626 cm^{-1} in chitin, at 1626 cm^{-1} in crystalline *N*-acetylglucosamine, and at 1623 cm^{-1} in crystalline *N*-acetylmannosamine) may represent one of two coupled vibrations of the unit cell, the other vibration being infrared inactive. The band at 1656 cm^{-1} in chitin may represent the uncoupled vibration in the vicinity of "missing" *N*-acetyl groups or in the noncrystalline or less regular regions of the sample.

This would certainly explain the variation of the ratio of intensities of the 1656 and 1626 cm^{-1} bands in chitin and the absence of the 1656 cm^{-1} band in chitan.

It was also observed that *N*-acetylglucosamine and *N*-acetylmannosamine in the vitreous form have a single broad amide I band at about 1640 cm^{-1} some 14 to 17 cm^{-1} higher than the corresponding sharp band in the spectra of the crystalline form of these compounds. The 1640 cm^{-1} band is probably due to uncoupled amide I vibrations in these compounds. The magnitude of the coupling constants observed in our compounds (30 cm^{-1} for chitin, 14 cm^{-1} for *N*-acetylglucosamine, and 17 cm^{-1} for *N*-acetylmannosamine) agrees with the values calculated for the amide I vibrations in polyglycine (18 cm^{-1}),[(135)] nylon (11 cm^{-1}),[(134)] and horsehair keratin (13 cm^{-1}).[(134)]

Whether the above explanation of the nature of the two amide I bands in chitin is correct, the two bands do appear to correspond to two distinct classes of amide groups in chitin, originating in distinct structural regions which may differ in their degree of order and which need not both contribute to the X-ray diffraction pattern of chitin.

It has been claimed that the frequency of the deformation mode of an anomeric axial (891 ± 7 cm^{-1}) or equatorial (844 ± 8 cm^{-1}) CH can be used to characterize the configuration at the anomeric centre of the glycopyranose residues.[(136–139)] Since chitan absorbs at 889 cm^{-1}, this empirical correlation would indicate that the sugar residues in this polysaccharide have a β-configuration at the anomeric centre. However, during deuterium exchange studies on chitan it was observed that progressive substitution of the hydrogen atoms of the hydroxyl groups by deuterium resulted in the gradual disappearance of the band at 889 cm^{-1}, with the simultaneous appearance of a band at 843 cm^{-1}. Similar results were also obtained on deuteration of chitin.

X-ray diffraction

The unit cell of chitan was determined[(140)] from the X-ray diffraction pattern of completely randomized, finely powdered chitan, using the method of Zsoldos. From the indexed diffraction patterns, the unit cell parameters were deduced as $a = 4{\cdot}80$, $b = 10{\cdot}32$, $c = 9{\cdot}83$ Å and $\beta = 112°$. Fibre diffraction patterns were recorded in order to check the unit cell parameters deduced from the powder ring analysis, to determine the space group of the monoclinic cell and to determine the structure factors of the observed reflections. The parameter b was found to be in agreement with the fibre repeat distance calculated from the layer line spacings of chitan fibres. It was concluded that there is a screw axis parallel to the fibre axis. As no further systematic absences were noted it was concluded that the space group is $P2_1$.

The density of chitan was determined by the flotation method using well-oriented bundles of chitan fibres, in a mixture of bromoform and benzene. It was found to be 1·495 g cm^{-3}. From the volume of the unit cell and density it was calculated that there are two residues of *N*-acetyl-glucosamine per unit cell of chitan. Since the crystallographic chain repeat is 10·32 Å and since from stereochemical criteria the β-(1→4) repeat distance of a single *N*-acetyl-D-glucosamine residue is 5·16 Å, there must be two residues in 10·32 Å. The two residues of *N*-acetylglucosamine in the unit cell belong to the same chain, and since there is a screw axis parallel to the fibre axis demanded by the space group, these two residues are crystallographic counterparts obtainable by a screw operation.

For the atomic coordinates the structure factors F_c were calculated for all possible reflections whose $\sin^2 \theta$ values were less than 0·13 because no reflections were recorded outside this range. The calculations were made using a computer programmed with the standard expression for structure factors for space groups $P2_1$. The X-ray intensities of all observed reflections from the fibre pattern were measured; the average integrated intensity from all four quadrants was calculated in each case, and the square root taken as observed structure factor F_0. Table 3.13 shows the *d*-values and *hkl* Miller indices of all observed reflections from the fibre pattern.

The macrostructure of unmodified chitan is markedly different from, and much more crystalline than, that of crustacean chitin.[140] There is no essential difference, however, between the unit cell proposed by Blackwell and that proposed by Dweltz *et al.* The independent studies of these authors confirmed the space group and parallel chain arrangement of the polymer, while emphasizing with instrumental data, the previously obtained information about partial deacetylation of the poly-*N*-acetyl-glucosamine.

Blackwell[141] in fact published X-ray diffraction patterns of pogonophore tube from *Oligobrachia ivanovi* after the deproteinized tube was bleached in Diaphanol, and compared the data with a spectrum of

TABLE 3.13. CALCULATED AND OBSERVED STRUCTURE FACTORS FOR CHITAN

hkl	$\sin^2\theta$ (calculated)	F_c	F_0	$\sin^2\theta$ (observed)	d (Å) (observed)
001	0·0072	22·81	22	0·007	9·12
101	0·0262	61·60	61	0·026	4·76
002	0·0286	3·13			
100	0·0300	88·00	83	0·030	4·45
102	0·0367	18·58	19	0·037	4·03
101	0·0481	0·39	V.F.	0·048	3·52
103	0·0615	1·51			
003	0·0644	8·73	V.F.	0·065	3·02
102	0·0806	5·02			
104	0·1006	10·89			
202	0·1047	15·46 }	M	0·105	2·38
201	0·1052	28·96 }			
004	0·1145	15·50			
203	0·1185	0·69			
200	0·1200	6·23			
103	0·1273	6·28			
010	0·0056	0			
011	0·0127	15·14	16	0·013	6·80
111	0·0318	25·09	26	0·032	4·31
012	0·0342	13·48	14	0·034	4·18
110	0·0356	7·00			
112	0·0422	9·18	V.F.	0·042	3·76
111	0·0537	10·14	V.F.	0·053	3·35
113	0·0670	16·57 }	W	0·068	2·96
013	0·0700	17·43 }			
112	0·0862	0·99			
114	0·1061	9·09 }			
212	0·1103	12·10 }	F	0·110	2·32
211	0·1108	7·85 }			
014	0·1200	16·23	F	0·120	2·23
213	0·1241	6·00			
210	0·1256	12·23			
020	0·0223	32·12	33	0·022	5·16
021	0·0295	16·65	17	0·030	4·45
121	0·0485	9·46	8	0·049	3·48
022	0·0509	15·05	16	0·051	3·41
120	0·0523	13·75 }	V.F.	0·057	3·23
122	0·0590	9·59 }			
121	0·0705	18·73	V.F.	0·070	2·91
123	0·0838	6·61			
023	0·0867	8·23			

TABLE 3.13 (*cont.*)

hkl	$\sin^2 \theta$ (calculated)	F_c	F_0	$\sin^2 \theta$ (observed)	*d* (Å) (observed)
122	0·1029	3·72			
124	0·1229	6·95			
222	0·1270	7·46			
221	0·1275	0·88			
030	0·0502	0			
031	0·0574	30·09	30	0·058	3·20
131	0·0764	22·47 }			
032	0·0788	30·90 }	36	0·077	2·78
130	0·0802	11·30 }			
132	0·0869	10·05	V.F.	0·089	2·58
131	0·0984	8·92			
133	0·1117	13·60	F	0·110	2·32
033	0·1146	7·58			
040	0·0893	13·43	F	0·089	2·58
041	0·0964	9·92	F	0·097	2·48
141	0·1155	8·75			
042	0·1179	8·62			
140	0·1193	4·44			
142	0·1259	10·42	F	0·126	2·17

V.F. = very faint.
F = faint.
(From M. Falk *et al.*, *Can. J. Chem.* **44,** 2269 (1966).)

Thalassiosira fluviatilis. Fom both his photographs he found a monoclinic unit cell, having dimensions $a = 4{\cdot}85$, $b = 10{\cdot}38$, $c = 9{\cdot}26$ Å and $\beta = 97{\cdot}5°$, and of course containing one chitin chain, in agreement with the parallel chain system proposed by Dweltz for β-chitin.

Further studies by the same author[114] indicated that the natural pogonophore tube consists of chitin with the same structure as that of the diaphanol-treated *Oligobrachia* plus another structural form giving rise to the extra reflections remarked in the *Zenkravichiana longissima* X-ray diffraction spectra. He proposed therefore that the single-form of β-chitin which occurs in the diaphanol-treated *Oligobrachia* tubes should be named β-chitin A. This form was shown to exist in three hydrate states (anhydrous, monohydrate and dihydrate structures) which occur separately and do not occur together in the same specimen. The pogonophore chitin also contains a second phase which is named β-chitin B. The latter has not been obtained in the absence of the A form.

On the basis of the X-ray diffraction results, the material from pogonophore tubes and from diatom spines is described as β-chitin A or chitan. *Loligo* pen β-chitin which contains the highest concentration of β-chitin B, also contains the highest proportion of deacetylated units. The hypothesis[114] was advanced that the β-chitin B structure might be adopted for partially deacetylated chains, while β-chitin A and chitan are synonymous of perfectly and homogeneously acetylated *N*-acetylglucosamine chains.

While nomenclature in the present situation is a matter of taste, it should be remarked that it is now clear that poly-*N*-acetylglucosamine occurs as such in nature, for instance in the diatom and pogonophora studied. Normally, when trying to obtain chitin from natural products, one notices that what is obtained is no longer the polymer, but a partially deacetylated polymer whose physico-chemical properties are altered with respect to the polymer isolated with no risk of deacetylation. It is thought that partial deacetylation can be present when the polymer still belongs to a living organism and it is combined with other molecules, and it is sure that the current harsh isolation treatment can introduce a further deacetylation.

Moreover, different crystal structures have been elucidated with the refined modern techniques, and these structures influence the possibilities that the polymer has to combine with other molecules.

ASSOCIATION OF CHITIN WITH PROTEIN

Early studies on orientation of fibrils in membranes and on protein orientation in chitinous structures pointed out that in insect cuticles, oriented poly-*N*-acetylglucosamine is associated with oriented tanned proteins.[142, 143]

The evidence of this point was that the chitin crystallites of insect cuticle were very small in dimension and on removing protein, large crystallites of chitin were formed. There is repetition of protein along the fibre axis of the chitin at 31 or 41 Å which corresponds to six or eight sugar residues.[144]

Neville[11] has demonstrated that a circadian rhythm takes place in the deposition of locust cuticle. In the tibia fibrous cuticle is laid down during the day with the fibre direction parallel to the axis of the tibia. The 31 Å structure is destroyed on removing the protein in 5% KOH. It is however very stable to steaming, either 2 min or 20 min steaming causing very little change in X-ray diffraction patterns.

These phenomena were later illustrated by Rudall,[144] who presented X-ray diffraction patterns and electron micrographs which indicated that there is a repeating structure of corpuscles regularly packed in rows along the chitin fibrils. Blake[145] illustrated the fitting of a chain of six *N*-acetylglucosamine residues into the cleft of the lysozyme molecule and outlined the numerous contacts or bonds between the protein and chitin. The said hexamer, reacting with protein is not a fully extended chain, as the successive residues are twisted relative to one another, while residue four is distorted from the normal "chair" form and it is even supposed that the polysaccharide is cleaved in this position. The insect cuticle suggests that there is an exact fitting of a protein with six residues of a chitin chain. The polymer molecule can afford a suitable position and stereoconfiguration for bonding with protein, and the closest packing of protein corpuscles along the chitin chain would be every 31 Å.

As far as fungi are concerned, the cell walls are relatively non plastic and no real success in obtaining parallel arrangements of chitin chains has been achieved. There is also the usual presence of glucans which are harder to remove than associate protein in animal chitin.[144]

It is known that in the cell walls of the yeast *Saccharomyces cerevisiae* not all the glucosamine is converted to chitin, the total glucosamine content of the wall being about 1 per cent. It is also known that chitin is localized in the sites of the bud scars. Studies[146] based upon the work by Bacon *et al.* contribute to the solution of the following questions:

(1) Is chitin present in cell walls without bud scars?
(2) What is the structural arrangement of chitin in bud scars?

In order to find answers to the above questions the amount of glucosamine contained in the walls of cells was determined (according to the number of bud scars) using both whole and extracted walls. Fractions obtained according to the method of Bacon *et al.* [147] (A) and after a subsequent 2 hr extraction with 2% HCl (B), as well as after a final 12 hr extraction with 5% KOH (C), were studied using i.r. spectroscopy, X-ray spectroscopy and electron microscopy.

TABLE 3.14. CHITIN BALANCE IN CELL WALLS OF THE YEAST *Saccharomyces cerevisiae* OF VARIOUS AGES

	Cells multiscarred	Non-fractionated population	Cells without bud scars
Distribution of particular cells	0 1–3 4→ 27% 38% 35%	0 1–3 4→ 57% 30% 13%	1→ 2%
Glucosamine of the entire cell wall (%)	1·4	0·9	0·3
Glucosamine in residue after Bacon's extraction and KOH (%)	7·1	5·7	2·6
Ratio of glucosamine in fraction after Bacon's extraction and KOH to total glucosamine (%)	52	30	23·6
Ratio of glucosamine in soluble fraction to total glucosamine	48	70	76
Glucosamine content in fraction following Bacon's extraction and KOH (%)	10·2	6·7	2·7

(From K. Beran *et al.*, *Proc. 2nd Intl. Symp. Yeast Protoplast*, Brno, (1968).)

The results presented in Table 3.14 indicate that the total glucosamine in cell walls, as well as its insoluble part in the wall fractions, increases with the number of scars. This agrees with the finding that chitin is localized in bud scars. Yet the values found for the walls and their fractions of scarless cells also indicate the presence of chitin. This is demonstrable by glucosamine present in an insoluble form, comprising 23·6% of the total. Although there is no direct evidence of its localization, one may presume that the chitin in these cells is localized in the birth scars.[148]

Results of i.r. spectroscopy and X-ray spectroscopy of the above wall fractions show a certain apparent inconsistency: whereas the fraction prepared according to Bacon already showed, in i.r. spectrum, the presence of all the typical chitin absorption zones, X-ray spectroscopy showed characteristic peaks only in fractions obtained after the final KOH extraction (C). Electron microscopic pictures of fraction (A) showed wall residues consisting of bud scars and their annuli surrounding the scar;[147] fractions (B) showed an identical structure, yet more granular. In fraction

(C) we can observe fusion of the material and partial disintegration of scar walls. Above results lead to the hypothesis that fibrils of changed chitin are present already in fraction (A), yet that they are mutually isolated by glucan structures. By further extractions of fractions (B) and (C) convergence of the fibrils is made possible resulting in the formation of the crystal grating responsible for the new X-ray spectrum. Identical presumptions correspond with the results of the studies by Rudall[144] obtained for the chitin–protein complex of insect cuticles.

Chitin seems to be localized not only in the bud scars, but also in the adjacent residual structure. Here it acts as stabilizer of glucan, which is therefore difficult to extract. The X-ray diffraction spectra in Fig. 3.12 strictly correspond to spectra taken on crab chitin in Fig. 3.13.[149]

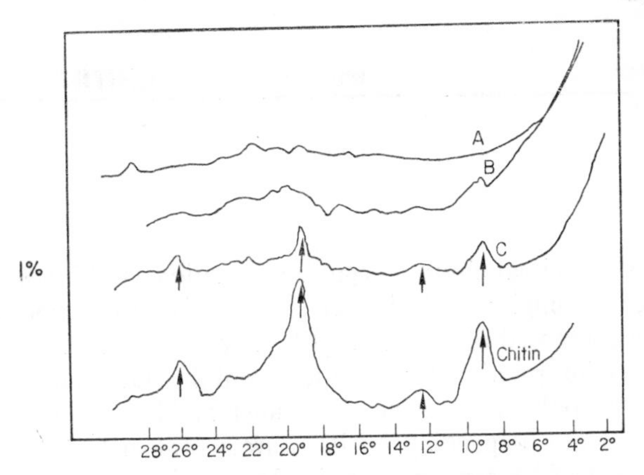

FIG. 3.12. Comparison of X-ray spectra of *Saccharomyces cerevisiae* yeast cell walls after extraction with chitin spectrum. Abscissa—degree of division; ordinate—% of transmission. (From K. Beran *et al.*, *Proc. 2nd Int. Symp. Yeast Protoplasts*, Brno (1968).)

Bacon, in his isolation of bud scars, found such regions to contain 50% chitins, and 30% of another glucan. The latter are predominantly $\beta(1\rightarrow6)$ or $\beta(1\rightarrow3)$ glucans. X-ray diffraction studies showed that fungi chitin is α-chitin, while crustacean chitin is β-chitin. It should be recalled that Kreger[35] found chitosan in the sporangiofores, and therefore it seems possible that there is a certain number of deacetylated residues along the chitin chains which cause imperfections in this α crystalline structure.

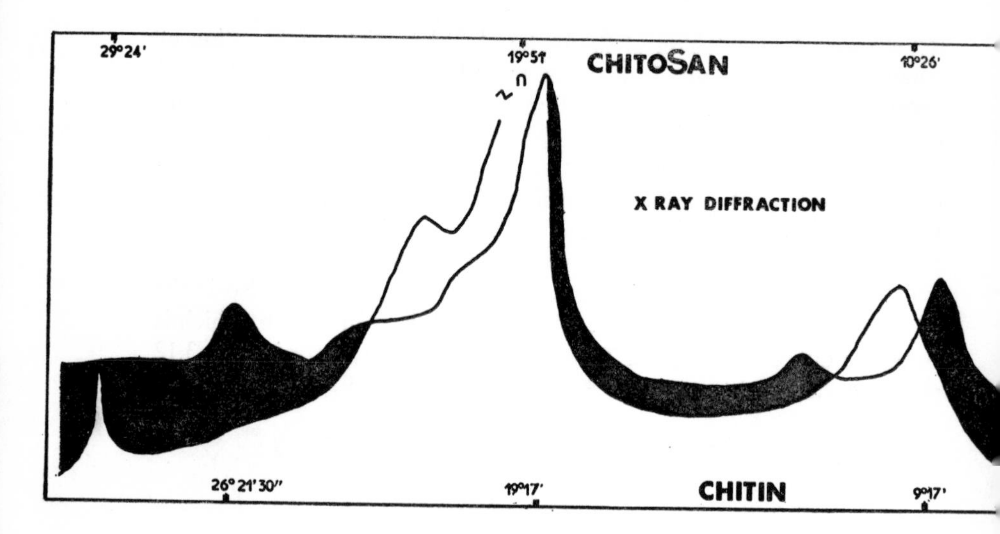

FIG. 3.13.

Whereas in animal chitin a regular association of chitin and protein takes place, in fungi the principal closely associated substance is a polysaccharide composed of glucose in $\beta(1 \rightarrow 3)$ and $\beta(1 \rightarrow 6)$ linkages.

The above information can be of use from the standpoint of chemical purity of the polymers. In fact two different kinds of molecules can accompany chitin depending on its source: protein for animal chitin, and polysaccharides for fungi chitin. In practice, should any protein accompany the chitin powders used in chromatography, one could think that proteins are partially responsible for metal ion collection.

Actually, while little can be done for purification of fungi chitin from polysaccharides, for a cheap and convenient preparation it seems that proteins can be removed from animal chitin by 5% KOH (which destroys the 31 Å structure) or by milder methods.[144]

Amino acids tentatively identified in the hydrolysate of the alkali-treated maggot cuticle were histidine, lysine, ornithine, phenylalanine, tyrosine, leucine, isoleucine, methionine, valine, alanine, glycine, glutamic acid, serine, threonine, and aspartic acid.[150] In the *Loligo* pen hydrolysate were detected histidine, lysine, ornithine, phenylalanine, tyrosine, methionine, and valine. All of them were at the trace level, especially compared with the high amount of glucosamine. In contrast with the findings of Hackman histidine and aspartic acids were found to be present at the trace level.

Many of the amino sugar links found in the protein-polysaccharides are loosened by alkali, and would be surely sensitive to the treatment based on 1 N sodium hydroxide solutions, unless they were sterically protected.

Therefore, while chitin may contain minor quantities of amino acids, it can be taken for granted that chitosan powders are free from amino acids, because of the strong alkali treatment to which chitin is submitted for chitosan preparation. In any case, chitin is generally prepared from those parts of the exoskeletons which are known to carry little protein, in order to reduce their presence in the product.

LIQUID CRYSTALS

Liquid crystal systems including fibrillar cellulose and partially deacetylated chitin have been reported.[151] A suspension of crystallite particles of chitin was prepared by treating 20 g of purified chitin from crab shells for 1 hr in 750 ml of 2·5 N hydrochloric acid under reflux. The excess acid was decanted and distilled water was added. The chitin hydrolysate was still mostly a sediment, and it was homogenized when still acid. From this treatment a stable isotropic suspension was obtained and the pH raised to 3·5 due to unacetylated amino groups at the crystallite surface. This suspension and a similar one of cellulose were the starting materials for the preparation of the liquid crystals as described below. The concentration was always less than 1%.

The formation of a permanently birefringent gel was first observed when a suspension of cellulose crystallites was heated on a steam bath. A soft, reddish-brown gel, formed on the surface of the heated suspension. It was found to be birefringent but without extinction directions, as does a powder of a birefringent crystal.

In the birefringent gel, low angle X-ray measurements have shown that the interparticle distance varies as the square root of the concentration. For a 15% gel it is about 400 Å. The properties of this system were found to be similar to the well-known behaviour of tobacco mosaic virus particles.

DERIVATIVES OF CHITIN

Chitin is insoluble in the usual solvents used for cellulose, it swells slightly in basic solvents and does not swell at all in the media currently used for esterification. Its derivatives therefore are formed with difficulty and their preparation requires drastic chemical operations, which in certain cases may lead to degradation. The different properties of chitin with respect to cellulose, are not only due to the presence of the acetyl-amino group, but also to its submicroscopic structure.

Research carried out on the hydrophilic properties of chitin by Plisko and Danilov[152] demonstrates that chitin is less hygroscopic than mercerized cellulose fibres, but more than ramie fibers, as from data in Table 3.15. In fact it is possible to freeze out at —6 °C nearly all the absorbed

TABLE 3.15. HYGROSCOPIC CHARACTERISTICS OF CHITIN AND CELLULOSIC FIBRES

Polymer	Hygroscopicity in relative humidity, %		Frozen water*
	65	100	
Chitin	8·9	34·7	1·4
Ramie fibres	5·5	19·7	12·5
Mercerized cellulose fibres (wood viscose)	8·9	32·6	26·2

* Amount of water remaining in the polymer after freezing at −6 °C, after saturation in relative humidity 100%, expressed in per cent of dry polymer.

(From E. A. Plisko *et al.*, *Khim. I Obmen Uglevodov*, 141 (1963).)

water of chitin, while the ramie and cellulose fibres retain considerable amounts of water. This points to the fact that chitin has a lower superficial activity and is less accessible to the infiltration of water and other reagents.

The heterogeneous acetylation of chitin proceeds slowly: completely acetylated chitin was obtained at 120 °C in the presence of fuming HCl.[153] A completely acetylated product was also obtained by bubbling hydrogen chloride through acetic anhydride.[154, 155]

In homogeneous acetylation conditions, the reaction is rather easier, but it does not reach a limiting value, as from data in Table 3.16.[152]

TABLE 3.16

Sample	Reaction temp. °C	Time hr	Catal. $HClO_4$ %	CH_3CO %	Substit. degree per glucose unit	Solubility
1	45	2·0	=	19·9	1·20	All samples soluble in 50 % resorcinol, phenol, and partially in *m*-cresol
2	45	4·0	=	23·8	1·45	
3	70	0·5	=	25·9	1·68	
4	80	0·5	=	26·3	1·70	
5	45	0·5	1·0	22·6	1·40	
6	45	4·0	1·0	24·3	1·50	
7	45	1·0	1·0	24·7	1·55	
8	75	2·0	1·0	25·5	1·60	
9	75	4·0	1·0	27·5	1·80	

(From E. A. Plisko *et al.*, *Khim. I Obmen Uglevodov*, 141 (1963).)

The acetate was soluble in formic acid and 50% resorcinol, but insoluble in other organic solvents. It dissolved with decomposition in concentrated sulphuric or hydrochloric acids. A chitin acetate containing 2·5 acetyl groups per sugar ring was strongly degraded in 90% formic acid.[46]

Chitin is degraded by the sulphuric acid in the mixtures used for nitrating cellulose, but chitin nitrates carrying 1·5 nitrate group per sugar ring have been prepared by using concentrated acid.[154] The product, which was partially soluble in formic acid, was stable but it ignited at 163 °C. Fuming nitric acid is also suitable for preparation of chitin nitrates which are not homogeneous as they have relatively short chains; they can be fractionated with formic or acetic acids.[98]

Chitin sulphates have been studied in order to prepare blood anticoagulants. Chlorosulphonic acid in pyridine or in 1,2-dichloroethane are suitable sulphating agents. Chitin disulphates show 20% of the anticoagulant potency of heparin with slight toxicity. Sulphated chitins have also been prepared by using sulphur trioxide in pyridine, dioxan, *N*,*N*-dimethylaniline, and bis-(2-chloroethyl)-ether; they are thickeners, adhesives and drilling muds.[156–163]

Ethers

It was established that by freezing an alkaline suspension, chitin swells, and upon repeated freezing, chitin dissolves. In so doing, a partial hydrolysis of acetyl groups occurs, but the chitin dissolution is mostly due to the weakening of the structure: in fact the regenerated chitin is insoluble at room temperature.[152] By this way, swelling of chitin can be enhanced and the diffusion of the reagents towards the reaction points of the macromolecule can be favoured. This technique was applied in the production of simple ethers of chitin.

Ethers of glycerine and chitin have been obtained by reacting glycerine monohydrochloride, or glycerine with alkali chitin: the products obtained with glycerine monohydrochloride are insoluble in alkali, while the others are soluble in NaOH 4–8%.[164]

Oxyethyl and carboxymethyl ethers have also been produced: the latter following the reaction:[165, 166]

$$[C_6H_7O_2(NHCOCH_3)\,(OH)_2]_n \xrightarrow[NaOH]{ClCH_2COOH}$$

$$\longrightarrow [C_6H_7O_2(NHCOCH_3)\,(OH)_{2-x}\,(OCH_2COONa)_x]_n$$

In Table 3.17 the conditions for the synthesis are reported. The raising of the reaction temperature reduces the portion of product soluble in water, but does not affect the substitution degree of the carboxymethylchitin. Increasing alkali concentration allows better solubility in water as a larger portion of the product is water-soluble, while higher substitution degree improves solubility too. Boiling carboxymethyl chitin lowers viscosity while coagulation is not observable.

According to other authors,[167] for the preparation of a partially *O*-carboxymethylated chitin 10 g of finely powdered chitin were soaked in 60 g of a 42% sodium hydroxide solution and then kept overnight *in vacuo*. Crushed ice was added to this mixture under stirring, until the volume reached 1 l. A solution of 300 ml of 4 N sodium chloroacetate was added to the paste thus obtained, after which the stirring was continued with no more cooling. In order to acetylate the partially deacetyl-

TABLE 3.17. CONDITIONS FOR SYNTHESIS, AND CHARACTERISTICS OF CARBOXYMETHYL CHITIN

NaOH concn %*	Moles NaOH	Moles Mono-chloro-acetic acid	Time hr	Temp. °C	Water-soluble portion %	NaOH %	Substit. degree per glucose unit	Viscosity of the 1% aq. soln. c. poise
40·0	4	2	20	20	57·5	6·90	0·84	16·63
				40	68·5	6·90	0·84	13·64
				60	40·7	6·44	0·80	7·41
	6	3	20	20	82·6	7·30	0·89	33·57
				40	69·0	7·51	0·92	30·18
				60	40·0	7·16	0·88	24·10
49·2	6	3	20	20	93·0	7·93	0·97	22·18
				40	76·0	8·76	1·08	11·13
				60	60·0	7·21	0·88	26·06
49·7	6	3	18·5	40	88·2	7·25	0·88	22·84
			12·0	60	88·8	6·13	0·75	30·20

* Chitin was impregnated in excess alkali solution.

(From E. A. Plisko *et al.*, *Khim. I Obmen Uglevodov*, 141 (1963).)

ated amino groups, 100 ml of acetic anhydride were added, while preventing the pH from going below 10 by the simultaneous addition of sodium hydroxide. After 24 hr stirring the mixture was dialysed against water for 3 days and against deionized water for 24 hr. The supernatant was lyophilized. 8·5 g of the sodium salt of partially *O*-carboxymethylated chitin were obtained, containing 14% water.

Wolfrom and coworkers[168] have treated chitosan with dimethyl sulphate and sodium hydroxide, and after acetylation and further manipulation obtained methylated chitin with a metoxyl content of 92%; the use of chitosan was justified with the difficulties in the etherification process mentioned above.

Chitin itself however, can be directly ethylated,[152] and results are presented in Table 3.18. These ethers are soluble in certain organic solvents and yield viscous solutions useful in producing membranes with good mechanical characteristics (5·5–7·3 kg mm^{-2} tensile strength and 5–23%

TABLE 3.18. CONDITIONS FOR ETHYLATION, AND CHARACTERISTICS OF THE ETHERS OF CHITIN

Sample	Moles		Regimen	C_2H_5O %	Remarks
	NaOH	C_2H_5Cl			
1	14	12	60 °C 1 hr	26·50	Soluble in benzene up to 40%.
2	16	15	80 °C 1 hr	27·10	Soluble in toluene, benzene
3*	16	15	130 °C 7 hr	33·30	Soluble in ketones, hydrocarbons and mixed solvents

* Reaction carried out under pressure.

(From E. A. Plisko *et al.*, *Khim. I Obmen Uglevodov*, 141 (1963).)

elongation.[165] As a point of difference between ethylcellulose and ethylchitin, the latter does not melt.[152]

No information on the interactions of metal ions with these derivatives is available.

Chitin gives a derivative containing one sodium atom per sugar unit when treated with sodium in liquid ammonia.[94] An alkali chitin was also reported after steeping the polymer in a 50% aqueous sodium hydroxide solution for 2 hr at 25 °C.[169, 170] This product gave with carbon disulphide a chitin xantate,[170] from which chitin was regenerated by acidification.

REFERENCES

1. R. A. A. MUZZARELLI and O. TUBERTINI, *Talanta* **16,** 1571 (1969); see also U.S. Patent 3, 635, 818 (1972), Italian priority 6 December 1968.
2. L. F. LELOIR, C. E. CARDINI and E. CABIB, *Comparative Biochemistry* (M. FLORKIN and H. S. MASON Eds.), Vol. 2, p. 97, Academic Press, London 1960.
3. H. BOSTROM and L. RODÈN, Metabolism of glycosaminoglycans, in *The Amino Sugars* (E. A. BALÁZS and R. W. JEANLOZ, Eds.), Vol. 2-B, Academic Press, London, 1966.
4. M. G. M. PRYOR, *Comparative Biochemistry* (M. FLORKIN and H. S. MASON, Eds.), Vol. 4, p. 371, Academic Press, London 1962.
5. R. DENNEL, *The Physiology of Crustacea* (T. H. WATERMAN, Ed.), Vol. 1, p. 447, Academic Press, London 1960.

6. K. M. Rudall, *Adv. Insect Physiol.* **1,** 257 (1963).
7. A. G. Richards, *Ann. Entomol. Soc. Am.* **40,** 227 (1947).
8. A. G. Richards and L. Pipa, *Smithsonian Inst. Misc. Coll.*, 137, 247 (1959).
9. S. Varagnolo, *Arch. Oceanogr. Limnol.* **15**-S, 83 (1968).
10. M. Lafon, *Bull. Inst. Oceanogr.* **45,** 1 (1948).
11. A. C. Neville, *Adv. Insect. Physiol.* **4,** 213 (1967).
12. C. H. Giles, A. S. A. Hassan, M. Laidlaw and R. V. R. Subramanian, *J. Soc. Dyers Colour.* **74,** 647 (1958).
13. R. H. Marchessault, F. F. Morehead and N. M. Walter, *Nature, Lond.* **184,** 632 (1959).
14. M. G. M. Pryor, *Proc. R. Soc.* B. 128, 393 (1940).
15. H. H. Scudamore, *Physiol. Zool.* **20,** 187 (1947).
16. D. F. Travis and U. Friberg, *J. Ultrastruct. Res.* **9,** 285 (1963).
17. M. Lafon, *Ann. Sci. Nat. Ser. Bot. Zool.* **11,** 113 (1943).
18. J. Brisou, C. Tysset, Y. de Rautlin de la Roy, R. Curcier, and R. Moreau, *Ann. Inst. Pasteur* **106,** 469 (1964).
19. J. H. Martin, *Limnol. Oceanogr.* **15,** 756 (1970).
20. B. J. Szabo, *Geochim. Cosmochim. Acta* **31,** 1321 (1967).
21. B. J. Szabo, *Carib. J. Sci.* **8,** 185 (1968).
22. H. J. M. Bowen, *Trace Elements in Biochemistry*, p. 241, Academic Press, 1966.
23. M. Tatsumoto and C. C. Patterson, The concentration of common lead in sea-water, in *Earth Science and Meteorites* (J. Geiss and E. D. Goldberg, Eds.), North-Holland, 1963.
24. J. F. Slowey and D. W. Hood, *Geochim. Cosmochim. Acta* **35,** 121 (1971).
25. D. W. Spencer and P. G. Brewer, *Geochim. Cosmochim. Acta* **33,** 325 (1969).
26. S. M. Manskaya and T. V. Drozdova, *Geochemistry of Organic Substances*, Pergamon, Oxford, 1968.
27. A. P. Vinogradov, *Elementary Chemical Composition of Marine Organisms*, A.N. USSR, Moscow, 1944.
28. V. A. Uspenskii, *Chimyia Tvedovo Topliva* **9,** 7 (1938).
29. S. M. Manskaya and T. V. Drozdova, *Geochimyia* 11 (1962).
30. L. J. Soin, *J. Paleontol.* **32,** 730 (1958).
31. E. Abderhalden and K. Heyns, *Bioch. Z.* **259,** 320 (1933).
32. N. Sharon, *The Amino Sugars* (E. A. Balázs and R. W. Jeanloz, Eds.) Vol. 2-A, p. 2, Academic Press, London, 1965.
33. P. A. Roelofsen and I. Hoette, Antonie van Leeuwenhoek, *J. Micribiol. Serol.* **17,** 297 (1951).
34. R. Frey, *Ber. Schweiz. Botan. Ges.* **60,** 199 (1950).
35. D. R. Kreger, *Biochim. Biophys. Acta* **13,** 1 (1954).
36. L. M. Carbonell and F. Kanetsura and F. Gil, *J. Bacteriology* **101,** 636 (1970).
37. J. M. Aronson and L. Machlis, *Am. J. Botany* **46,** 292 (1959).
38. H. J. Blumenthal and S. Roseman, *J. Bacteriol.* **74,** 222 (1957).
39. L. Zechmeister and G. Tóth, *Z. Physiol. Chem.* **223,** 53 (1934).
40. P. Karrer and A. Hoffmann, *Helv. Chim. Acta* **12,** 616 (1929).
41. P. Karrer and G. von François, *Helv. Chim. Acta* **12,** 986 (1929).

42. K. Heller, L. Claus and J. Huber, *Z. Naturforsch.* **14**-B, 476 (1959).
43. G. van Iterson, K. H. Meyer and W. Lotmar, *Rev. Trav. Chim.* **55,** 61 (1936).
44. A. N. J. Heyn, *Protoplasma* **25,** 372 (1936).
45. J. M. Diehl and G. van Iterson, *Kolloid Z.* **73,** 142 (1935).
46. K. H. Meyer and H. Wherli, *Helv. chim. Acta* **20,** 353 (1937).
47. M. V. Tracey, *Modern Methods of Plant Analysis* (K. Paech and M. V. Tracey Eds.), Vol. 2, p. 264, Springer, Berlin, 1955.
48. F. L. Campbell, *Ann. Entomol. Soc. Am.* **22,** 401 (1929).
49. H. Brunswik, *Biochem. Z.* **113,** 111 (1921).
50. J. S. D. Bacon, E. D. Davidson, D. Jones and I. F. Taylor, *Biochem. J.* **101,** 36-C (1966).
51. A. G. Richards, *The Integument of Arthropods*, University of Minnesota Press, Minneapolis, 1951.
52. N. W. Runham, *J. Histochem. Cytochem.* **10,** 504 (1962).
53. N. W. Runham, *J. Histochem. Cytochem.* **9,** 87 (1961).
54. T. N. Salthouse, *J. Histochem. Cytochem.* **10,** 109 (1962).
55. L. Zechmeister and G. Tóth, *Chem. Ber.* **64,** 2028 (1931).
56. F. Zilliken, G. A. Braun, C. S. Rose and P. György, *J. Am. Chem. Soc.* **77,** 1296 (1955).
57. S. A. Barker, A. B. Foster, M. Stacey and J. M. Webber, *Chem. Ind.* 208 (1957).
58. S. A. Barker, A. B. Foster, M. Stacey and J. M. Webber, *J. Chem. Soc.* 2218 (1958).
59. M. S. Fuller and I. Barshad, *Am. J. Botany* **47,** 105 (1960).
60. J. M. Aronson and R. D. Preston, *Proc. R. Soc.* **152**-B, 346 (1960).
61. E. J. Winkler, L. A. Douglas and D. Pramer, *Biochim. Biophys. Acta* **45,** 393 (1960).
62. G. A. Paszkiewicz, J. Niedzwicka and J. Popowicz, *Chem. Anal.* **16,** 443 (1971).
63. K. Beran and J. Rehacek, *Proc. 2nd Int. Symp. Yeast Protoplasts*, Brno, 1968; see also *Folia Microbiologica*, to be published in 1973.
64. L. Rzucidlo, *Postepy Mikrobiol.* **4,** 57 (1965).
65. J. N. BeMiller, *Methods of Carbohydrate Chemistry*, Vol. 5, p. 103, 1965.
66. G. W. Rigby, U.S. Patent 2, 040, 879 (1936)
67. R. Blumberg, C.L. Southall, N.J. van Rensburg and O.B. Volckman, *J. Sci. Food Agr.* **2,** 571 (1951).
68. S. A. Barker, A. B. Foster, M. Stacey and J. M. Webber,
69. R. H. Hackman, *Australian J. Biol. Sci.* **7,** 168 (1954).
70. R. S. Whistler and J. N. BeMiller, *J. Org. Chem.* **27,** 1161 (1962).
71. S. T. Horowitz, S. Roseman and H. J. Blumenthal, *J. Am. Chem. Soc.* **79,** 5046 (1957).
72. A. B. Foster and R. H. Hackman, *Nature, Lond.* **180,** 40 (1957).
73. M. Takeda and E. Abe, *Norin sho Suisan Koshuso Kenkyu Hokoku* **11,** 399 (1962).
74. M. Takeda and H. Katsuura, *Suisan Daigaku Hokoku* **13,** 109 (1964).
75. P. Broussignac, *Chim. Ind. Génie Chim.* **99,** 1241 (1968).

76. H. J. EINBRODT and W. STOBER, *Naturwissenschaften* **47,** 84 (1960).
77. N. OKAFOR, *Biochim. Biophys. Acta* **101,** 193 (1965).
78. E. PLESSMANN CAMARGO, C. P. DIETRICH, D. SONNEBORN and J. L. STROMINGER, *J. Biol. Chem.* **242,** 3121 (1967).
79. J. S. LOVETT and E. C. CANTINO, *Am. J. Bota.* **47,** 550 (1960).
80. L. GLASER and D. H. BROWN, *J. Biol. Chem.* **228,** 729 (1957).
81. C. A. PORTER and E. G. JAWORSKI, *Biochemistry* **5,** 1149 (1966).
82. E. G. JAWORSKI, L. C. WANG and W. D. CARPENTER, *Phytopathology* **55,** 1309 (1965).
83. F. G. CAREY, *Comp. Biochem. Physiol.* **16,** 155 (1965).
84. C. A. PORTER and E. G. JAWORSKI, *J. Insect Physiol.* **11,** 1151 (1965).
85. L. GLASER and D. H. BROWN, *Biochim. Biophys. Acta* **23,** 449 (1957).
86. P. W. KENT and M. R. LUNT, *Biochim. Biophys. Acta* **28,** 657 (1958).
87. M. R. LUNT and P. W. KENT, *Biochim. J.* **78,** 128 (1961).
88. F. G. CAREY and G. R. WYATT, *Biochim. Biophys. Acta* **41,** 178 (1960).
89. D. J. CANDY and B. A. KILBY, *J. exp. Biol.* **39,** 129 (1962).
90. H. ZALUSKA, *Acta Biol. Exp.* **19,** 339 (1960).
91. V. R. MEENAKSHI and B. T. SCHEER, *Comp. biochem. Physiol.* **3,** 30 (1961).
92. M. L. BADE and G. R. WRIGHT, *Biochem. J.* **83,** 470 (1962).
93. J. S. BRIMACOMBE and J. M. WEBBER, *Mucopolysaccharides*, Elsevier, Amsterdam, 1964, p. 18.
94. *Nature* **234,** 16 (1971).
95. R. H. HACKMAN and M. GOLDBERG, *Australian J. biol. Sci.* **18,** 953 (1965).
96. R. IKAN, *Natural Products*, p. 67, Academic Press, London, 1969.
97. R. JEANLOZ and E. FORCHIELLI, *Helv. Chim. Acta* **33,** 1690 (1950).
98. G. L. CLARK and A. L. SMITH, *J. Phys. Chem.* **40,** 863 (1936). nitrate
99. E. R. PURCHASE and C. E. BROWN, *Organic Synthesis* **26,** 36 (1946).
100. R. F. POWNING and H. IRZYKIEWICZ, *J. Chromatogr.* **17,** 621 (1965).
101. D. R. WHITAKER, M. E. TATE and C. T. BISHOP, *Can. J. Chem.* **40,** 1885 (1962).
102. R. KUHN and G. KRUGER, *Chem. Ber.* **90,** 264 (1957).
103. P. P. VON WEIMARN, *Ind. Eng. Chem.* **19,** 109 (1927).
104. C. E. ZOBELL and S. C. RITTENBERG, *J. Bact.* **35,** 275 (1938).
105. L. CAMPBELL and O. B. WILLIAMS, *J. Gen. Microbiol.* **5,** 894 (1951).
106. D. W. LEAR, *Proc. Symp. Marine Microbiology*, Thomas, Springfield, Illinois, 1963.
107. D. F. WATERHOUSE, R. H. HACKMAN and W. J. MCKELLAR, *J. Insect Physiol.* **6,** 96 (1961).
108. C. JEUNIAUX, *Chitine et Chitinolyse*, Masson, Paris 1963.
109. C. JEUNIAUX, *Arch. Int. Physiol. Biochim.* **69,** 384 (1961).
110. M. V. TRACEY, *Biochem. J.* **61,** 579 (1955).
111. R. SPENCER, *Nature, Lond.* **190,** 938 (1961).
112. L. R. BERGER and R. S. WEISER, *Biochim. Biophys. Acta* **26,** 517 (1957).
113. T. MIYAZAKI and Y. MATSUSHIMA, *Bull. Chem. Soc. Japan* **41,** 2754 (1968). Have copy
114. J. BLACKWELL, *Biopolymers* **7,** 281 (1969).
115. N. F. DWELTZ, *Biochim. Biophys. Acta* **44,** 416 (1960).
116. K. H. MEYER and G. W. PANKOW, *Helv. Chim. Acta* **18,** 589 (1935).

117. K. H. MEYER and H. MARK, *Ber.* **61,** 1936 (1928).
118. K. H. GARDNER and J. BLACKWELL, *J. Polymer Sci.* **36–C** 327 (1971).
119. D. CARLSTROM, *J. Biophys. Biochem. Cytol.* **3,** 669 (1957).
120. F. G. PEARSON, R. H. MARCHESSAULT and C. Y. LIANG, *J. Polymer Sci.* **43,** 101 (1960).
121. R. H. MARCHESSAULT, F. G. PEARSON and C. Y. LIANG, *Biochim. Biophys. Acta* **45,** 499 (1960).
122. M. FALK, D. G. SMITH, J. MCLACHLAN and A. G. MCINNES, *Can. J. Chem.* **44,** 2269 (1966).
123. J. MCLACHLAN, A. G. MCINNES and M. FALK, *Can. J. Botany* **43,** 707 (1965).
124. A. BERGER and K. LINDERSTRØM-LANG, *Arch. Biochem. Biophys.* **69,** 106 (1957).
125. L. J. BELLAMY, *The Infrared Spectra of Complex Molecules*, Methuen, London, 1958.
126. C. H. BAMFORD, A. ELLIOTT and W. E. HANBY, *Synthetic Polypeptides*, p. 208, Academic Press, London, 1956.
127. N. L. JOHNSON and D. C. PHILLIPS, *Nature, Lond.* **202,** 588 (1964).
128. M. BEER, H. B. KEESLER and G. B. B. M. SUTHERLAND, *J. Chem. Phys.* **29,** 1097 (1958).
129. S. E. DARMON and K. M. RUDALL, *Disc. Faraday Soc.* **9,** 251 (1950).
130. K. BERAN, J. REHACEK and O. SEICHERTOVA, *Acta Fac. Med. Univ. Brun.* **37,** 171 (1970).
131. A. J. MICHELL and G. SCURFIELD, *Aust. J. Biol. Sci.* **23,** 354 (1970).
132. T. MIYAZAWA, *J. Chem. Phys.* **32,** 1647 (1960).
133. T. MIYAZAWA and E. R. BLOUT, *J. Am. Chem. Soc.* **83,** 712 (1961).
134. E. M. BRADBURY and A. ELLIOTT, *Polymer* **4,** 47 (1963).
135. S. KRIMM, *J. Mol. Biol.* **4,** 528 (1962).
136. R. E. MORENO, F. KANETSUNA and L. M. CARBONELL, *Arch. Biochem. Biophys.* **130,** 212 (1969).
137. S. A. BARKER, E. J. BOURNE, M. STACEY and D. H. WHIFFEN, *J. Chem. Soc.* 171. (1954).
138. A. B. FOSTER and J. M. WEBBER, *Adv. Carbohyd. Chem.* **15,** 361 (1960).
139. H. SPEDDING, *Adv. Carbohyd. Chem.* **19,** 23 (1964).
140. N. E. DWELTZ, J. ROSS COLVIN and A. G. MCINNES, *Can. J. Chem.* **46,** 1513 (1968).
141. J. BLACKWELL, K. D. PARKER and K. M. RUDALL, *J. Mol. Biol.* **28,** 383 (1967).
142. L. E. R. PICKEN, M. G. M. PRYOR and M. M. SWANN, *Nature, Lond.* **159,** 434 (1947).
143. L. E. R. PICKEN and W. LOTMAR, *Nature, Lond.* **165,** 599 (1950).
144. K. M. RUDALL, *J. Polymer. Sci.* **28 – C,** 83 (1969).
145. C. C. F. BLAKE, L. N. JOHNSON, G. A. MAIR, A. C. T. NORTH, D. C. PHILLIPS and V. R. SARMA, *Proc. R. Soc.* **167-B,** 429 (1967).
146. K. BERAN and J. REHÁCEK, *Antonie van Leeuwenhoek* **35-S,** 3 (1969).
147. J. S. D. BACON, E. D. DAVIDSON, D. JONES and I. F. TAYLOR, *Biochem. J.* **101,** 36 (1966).
148. K. BERAN, J. REHACEK and O. SEICHERTOVA, *Proc. 2nd Int. Symp. Yeast Protoplasts*, Brno, 1968.

149. R. A. A. Muzzarelli, *Rév. Int. Océanogr. Médicale* **21,** 93 (1971).
150. M. A. Attwood and H. Zola, *Comp. biochem. Physiol.* **20,** 993 (1967).
151. R. H. Marchessault, F. F. Morehead and N. M. Walter, *Nature, Lond.* **184,** 632 (1959).
152. E. A. Plisko and S. N. Danilov, *Khim. I Obmen Uglevodov*, 141 (1963).
153. P. Schorigin and E. V. Chwig, *Chem. Ber.* **68,** 271 (1935).
154. P. Schorigin and E. Hait, *Chem. Ber.* **67,** 1712 (1934).
155. P. Schorigin and E. Hait, *Chem. Ber.* **68,** 971 (1935).
156. S. Bergstrom, *Naturwissenschaften* **23,** 706 (1935).
157. S. Bergstrom, *Z. physiol. Chem.* **238,** 163 (1936).
158. I. B. Cushing, R. V. Davis, E. J. Kratovil and D. V. MacCorquodale, *J. Am. Chem. Soc.* **76,** 4590 (1954).
159. T. Astrup, I. Galsmar and M. Volkert, *Acta Physiol. Scand.* **8,** 215 (1944).
160. J. Piper, *Acta Pharmacol. Toxicol.* **2,** 138 (1946).
161. O. C. Barsøe and S. Selsø, *Acta Pharmacol. Toxicol.* **2,** 367 (1946).
162. L. W. Roth, I. M. Shepherd and R. K. Richards, *Proc. Soc. Exp. Biol. Med.* **86,** 315 (1954).
163. R. V. Jones, U.S. Patent 2,689,244 (1954).
164. S. N. Danilov and E. A. Plisko, *Z. Obsc. Khimii* **24,** 2071 (1954).
165. S. N. Danilov and E. A. Plisko, *Z. Obsc. Khimii* **28,** 2217 (1958).
166. S. N. Danilov and E. A. Plisko, *Z. Obsc. Khimii* **31,** 469 (1961).
167. T. Miyazaki and Y. Matsushima, *Bull. Chem. Soc. Japan* **41,** 2723 (1968).
168. M. L. Wolfrom, J. R. Vercellotti and D. Horton, *J. Org. Chem.* **28,** 278 (1963).
169. R. Senju and S. Okimasu, *Nippon Nogeikagaku Kaishi* **23,** 432 (1950).
170. C. J. B. Thor and W. F. Henderson, *Am. Dyestuff Rep.* **29,** 461 (1940).

CHAPTER 4

CHITOSAN

PRESENTATION

In 1894 Hoppe-Seyler fused potassium hydroxide at 180° C with chitin and obtained a product with diminished acetyl content, that he called chitosan.[1]

Actually, methods based on alkaline treatments should be employed for *N*-deacetylation, as *N*-acetyl groups cannot be removed by acidic reagents, as effectively as with alkaline treatments. However, it is known that partial deacetylation occurs under harsh treatment.

Chitosan was also studied by Araki,[2] and von Furth and Russo concluded that three out of four acetyl groups can be removed from chitin.[3]

Rigby in his patent on chitosan preparation presented a treatment of chitin with 40% aqueous solution of sodium hydroxide, for 4 hr at 110 °C.[4] Evidence of the chitosan identity was put forward by Clark who registered X-ray diffraction data.[5] However, this treatment should not be prolonged. Infrared spectrophotometry can be useful for checking the removal of acetyl groups.[6]

It should be kept in mind that chitin and chitosan are names that do not strictly refer to a fixed stoichiometry: in practice chitin is poly-*N*-acetylglucosamine (proposed name chitan) deacetylated a very little while chitosan is the same, deacetylated as far as possible, but not enough to be called polyglucosamine. As the deacetylation extent can vary, one sees that these are family names for two classes of compounds.

The most important preparations of chitosan (free base) are reported here with attention to its use for metal ion collection in chromatography and related works.

PREPARATIONS

Method of Horowitz[7, 8]

Thirty grams of chitin can be converted to chitosan by fusion with 150 g solid potassium hydroxide in a nickel crucible under stirring in a nitrogen atmosphere. After 30 min at 180 °C the melt is poured cautiously into ethanol and the precipitate is washed with water to neutrality. Purification can be done by dissolution in 5% formic acid and sodium hydroxide precipitation. A 95% removal of acetyl groups was claimed; however, the chain length after dialysis was found to have about twenty units only, which is not very much for a polymer to be used in chromatography.[9]

Method of Rigby and Wolfrom[4, 10, 11]

Fifty grams of chitin are treated with 2·4 l. of a 40% aqueous solution of sodium hydroxide at 115 °C for 6 hr under nitrogen. After cooling, the mixture is filtered and washed with water to neutral reaction. An 82% removal of acetyl groups can be obtained.

These procedures can be improved in order to reduce the yellowish colour formation and to avoid the cumbersome nitrogen atmosphere and nickel vessels.

Method of Broussignac[12]

A systematic survey by Broussignac led to the following deacetylation mixture which is nearly anhydrous: potassium hydroxide 50%, 96° ethanol 25% and monoethyleneglycol 25% by weight. To prepare this mixture the two solvents are first mixed and then potassium hydroxide is added under stirring in small portions. The dissolution is exothermic and the temperature can reach 90 °C during this step. This mixture can be used in both glass or stainless steel reactors. A stainless steel reactor consisting of a steam heating system, a stirrer and a reflux is at present in operation. It contains 360 kg of the deacetylation mixture, and 27 kg of chitin. The temperature is 120 °C and corresponds to the boiling of the mixture. Chitin should be dried before introduction into the reactor, of course. The treatment is continued for the desired length of time, and after filtration chitosan is washed with water until neutral reaction. Chitosan can be dried at a moderated temperature. In Fig. 4.1 viscosity and per cent

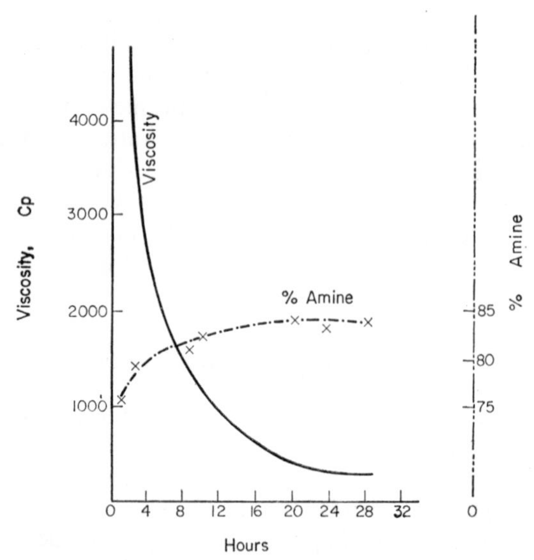

FIG. 4.1. Deacetylation of chitin and formation of free amino groups of chitosan vs. time. The viscosity of the chitosan obtained after deacetylation is also shown. (From P. Broussignac, *Chim. Ind. Gén. Chim.* **99**, 1241 (1968).)

of free amino groups are given vs. deacetylation treatment time. One can observe that the viscosity range of the obtained chitosan is very broad, and depends on the length of treatment. Amino groups are deacetylated to the extent of about 83% in 16 hr, which is very satisfactory. The Broussignac method yields 7% chitosan of the crab shells weight. The author has remarked that when doing filtration, the alkaline mixture contains certain quantities of chitosan in dissolved form. Upon dilution with water a white flocculent precipitate can be obtained from the spent deacetylation mixture. This amount of chitosan should be added to the

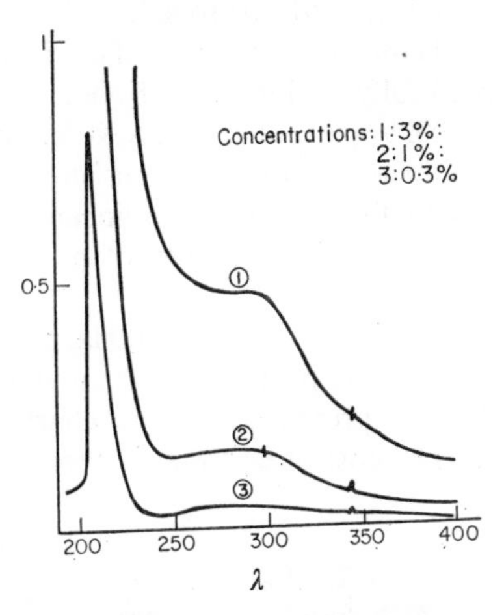

FIG. 4.2. Ultraviolet spectrum of chitosan. (From P. Broussignac, *Chim. Ind. Gén. Chim.* **99,** 1241 (1968).)

yield mentioned. In order to spare on ethanol and to lower the amount of chitosan dissolved in the deacetylation mixture, the latter has been modified by the author as follows.

Five hundred grams of KOH are dissolved in 450 ml of diethyleneglycol and 76 g of chitin are added. The mixture is kept at 170 °C for 6 hr.

Method of Fujita[13]

Ten parts of chitin are mixed with ten parts of 50% NaOH, kneaded, mixed with 100 parts of liquid paraffin, and stirred for 2 hr at 120 °C; then the mixture is poured into 80 parts of cold water, filtered and thoroughly washed with water; the yield is 8 parts of chitosan. The free amino group content is 0·92 per glucose residue. This method is simple and requires much less hydroxide than the other methods reported.

Industrial production

The above-reported production methods are good for small samples, but they are not suitable for commercial stocks. Previous methods of processing wastes from marine food factories gave consideration neither to recovery of the chemical used nor to sales of the derivatives and by-products that also can be made. As the role of chitin and chitosan in new markets has never been fully explored, the Food, Chemicals & Research Laboratories, Ltd., 4900, 9th Avenue NW, Seattle, installed a pilot plant to produce enough chitin and chitosan of consistent, controllable quality to permit researchers to find out new applications of these polymers, and to determine the economic feasibility of marketing those applications.

The pilot plant capacity is presently of 1000 kg per month; in addition, it produces an equal amount of protein powder, and it is linked to another process for producing fish proteins. The flow chart of the pilot plant is presented in Fig. 4.3. The cost of chitosan is about $4·00 per kg, but in

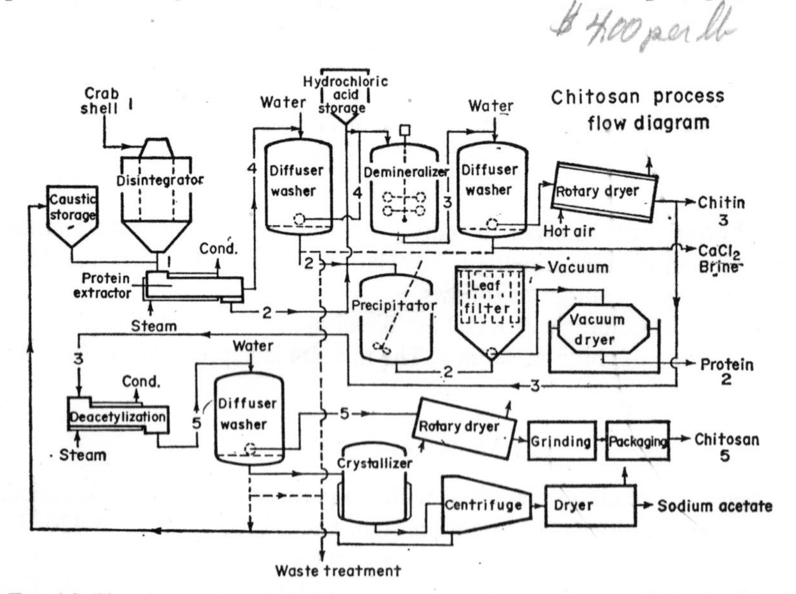

FIG. 4.3. Flowsheet of the pilot plant for the production of chitosan. (1) shell, (2) protein, (3) chitin, (4) deproteinized shell, (5) chitosan. (Courtesy of Food, Chemicals & Research Laboratories, Ltd., Seattle.)

full-scale commercial production this might be reduced to about \$1·00 per kg.

The effects of the experimental conditions upon the degradation of the polymer and its deacetylation degree have been examined recently by Russian authors,[14] who studied the preparation of chitosan according to Rigby's specifications and under inert atmosphere. Chitosan prepared under nitrogen yields solutions in 2% acetic acid with higher viscosity. A 0·5% chitosan solution in 2% acetic acid at pH = 3·6 has been used for fractionation studies; its viscosity was 12·80 and its deacetylation degree 0·79. The primary fractionation was carried out with acetone at pH = 6·5 and two fractions were collected: a secondary fractionation yielded nine fractions whose viscosities were further determined. The data are reported in Table 4.1. As the nitrogen content in the nine fractions is practically

TABLE 4.1. RESULTS OF THE FRACTIONATION OF CHITOSAN

Fraction number	Fraction %	$[\eta]$	Nitrogen content, %
1	5·30	15·10	8·67
2	8·51	13·20	
3	28·63	12·75	7·58
4	20·29	12·05	8·66
5	10·41	10·75	8·22
6	1·74	10·68	
7	8·94	8·20	8·72
8	3·45	3·12	
9	5·01	2·30	7·31

(From L. A. Nud'ga *et al.*, *Zh. Obshchei Khimii*, **41**, 2555 (1971).)

constant, it seems that the fractionation is mainly based on molecular weight differences. From these data the authors calculated that the chitosan so prepared was scarcely polydispersed, as expected (see page 156).

CHITOSAN CHARACTERISTICS

Solubility of chitosan and its salts

Chitosan is insoluble in water and in organic solvents; it is also insoluble in alkali and in mineral acids except under certain conditions. In the presence of a limited amount of acid it is soluble in water–methanol, water–ethanol, water–acetone and other mixtures. Chitosan is soluble in formic and acetic acids, and according to Peniston in 10% citric acid. Hamiter reported dissolution of chitosan in pyruvic and lactic acids. Even though a systematic survey was not reported, other organic acids fail to dissolve chitosan. Inorganic acids can dissolve chitosan at certain pH values after prolonged stirring and warming. Nitric acid can dissolve some chitosan, but some time after dissolution one can observe a jelly white precipitate. With cold concentrated nitric acid generally one can observe a partial collapse of the grains, but dissolution is not very fast. Hydrochloric acid also requires heating and stirring for hours. Sulphuric acid does not dissolve chitosan because it forms chitosan sulphate which is a white crystalline solid. Perchloric acid can dissolve chitosan easier.

The above statements are for a 100–200 mesh powder obtained according to the Broussignac treatment, but they do not hold for chitosan obtained from a precipitation operation; in fact, precipitated chitosan can be more easily brought into solution by hydrochloric, nitric and perchloric acids.

To prepare chitosan perchlorate,[10] chitosan was suspended in a mixture of 132 ml of glacial acetic acid and 33 ml of 60% perchloric acid. The suspension was stirred at 8 °C for 16 hr prior to filtration and drying with ether; the yield of chitosan perchlorate was 3·2 g. The product corresponded to the formula $C_6H_9O_4\,(NHCOCH_3)_{0\cdot15}\,(NH_3^+ClO_4^-)$.

From formic acid solutions, chitosan can be precipitated by 0·1 N sulphuric acid, by copper sulphate and ammonium sulphate, as white crystalline chitosan sulphate. Concentrated hydrochloric acid also precipitates chitosan. Sodium sulphite, sodium thiosulphate and triethanolamine also precipitate chitosan from formic or acetic acid solutions.

It is worth while to note that the precipitate is white when the precipitation is carried out with copper sulphate, i.e. copper ions are left in solution by chitosan sulphate, while the controlled precipitation of the free base chitosan, as described in a further chapter, leads to the collection of copper ions.

Recently the dissolution of cellulose and polyuronides by mixtures of dimethylformamide and dinitrogen tetroxide has been achieved,[15] and the applicability of this system toward chitosan and a group of marine polymer has been reported.[16] The solubilities of 1 g of polymer in 50 ml of dimethylformamide, with a dinitrogen tetroxide to polymer ratio of 3:1 are in Table 4.2.

TABLE 4.2. SOLUBILITIES OF MARINE POLYSACCHARIDES IN DIMETHYLFORMAMIDE + DINITROGEN TETROXIDE

Polysaccharide	Fraction dissolved
Alginic acid	70
Propylene glycol alginate	100
Agar	100
Chitin	5 (swells)
Chitosan	100
Laminaran	100
Mucilage of *Ulva lactuca*	75

(From G. G. Allan *et al.*, *Chem. Ind.* 127 (1971).)

Infrared spectral analysis of the starting materials and products indicated the absence of chemical modifications and that their solubilities in appropriate solvents were identical. The highly ionic polymers alginic acid, chitosan and the mucilage of *Ulva lactuca* gave non-viscous solutions presumably owing to tight chain coiling induced by common ion effects. On the contrary, the essentially non-ionic cellulose, agar and propylene-glycol alginate gave viscous solutions.

For the mechanism by which solution occurs, it was suggested that hydrogen bonds are disrupted by NO^+ ions.

Sulphonation and nitration of chitosan, yield products which are both salts and esters. Chitosan sulphate derivatives have been studied in connection with the anticoagulant activity of sulphated compounds, like heparin.[11] The presence of acid sensitive *N*-sulphate groups in the sulphated chitosan was demonstrated by observing the rate of evolution of nitrogen in the Van Slyke deamination apparatus: the sulphated chitosan reacts more rapidly under these conditions than does heparin, in accordance with general results. Sulphated chitosan has in fact high anticoagulant activity.[17–20]

A homogeneous sulphation of chitosan using a sulphur trioxide-*N*,*N*-dimethylformamide complex in an excess of the latter, has been reported.[11] The sulphur trioxide complex can be formulated as a dipolar ion, whose reaction with a nucleophilic group such as a primary amine should involve attack at the sulphur atom to effect sulphation on nitrogen or oxygen with the concomitant release of *N*, *N*-dimethylformamide.

$$RNH_2 + HC{=}N^+(CH_3)_2{-}O{-}SO_2{-}O^- \dashrightarrow RNHSO_2OH + HCON(CH_3)_2$$

The above procedure applied to chitosan, yielded an amorphous product containing one sulphoamino group and one sulphate acid ester group per monomer unit. Thirty ml of the sulphur trioxide-*N*,*N*-dimethylformamide were necessary for 2 g of chitosan. Chitosan dissolves in this mixture and after 12 hr the crude product was isolated as the sodium salt by addition of sodium bicarbonate. The precipitate was redissolved in water and dialysed for 3 days. Sulphation of chitosan can also be performed with chlorosulphonic acid and dry pyridine.[11]

Transformation of chitosan in concentrated sulphuric acid was investigated by Nagasawa.[21, 22] Finely powdered chitosan (average molecular weight 120,000; 92% free amine) was dried *in vacuo* over P_2O_5 for 3 hr at 75 °C. Two grams of the dried chitosan were added in small portions during 15 min to 40 ml of conc. sulphuric acid with vigorous stirring at the temperature indicated, until a gelatinous homogeneous state was reached. The reaction mixture was then poured into 400 ml of ether at constant temperature, and the precipitate was collected on a glass filter, and after washing it was dissolved in water and neutralized with cold 30% NaOH. The neutralized solution was dialysed against tap water for 40 hr. The average molecular weight of the sulphated chitosan was 31,000.

Neither reaction temperature (0–30 °C) nor time (1–10 hr) affected the degree of sulphation of chitosan, but it was markedly influenced by the temperature of the ether used to separate sulphated products. As previously reported by the same authors,[23] a remarkable depolymerization occurred when chitin was treated with conc. sulphuric acid at -5 °C for 2 hr, and resulted in yielding the non-dialysable (50%) and the dialysable (44%) fractions even after treatment at 0 °C for 10 hr. This extreme stability to depolymerization was attributed to the stabilizing effect of the free amino groups in the chitosan molecule. In order to clarify a relationship between the residual *N*-acetyl content and depolymerization of chitosan, partially deacetylated chitin samples were treated with conc. sulphuric acid: the larger the increase of residual *N*-acetyl content, the greater the degree of polymerization. As expected, chitosan was greatly depolymerized by treatment with hydrated conc. sulphuric acid, and sulphated to a lesser extent than with conc. sulphuric acid.

A hydrolysis constant $K_i = 7{\cdot}0 \times 10^{-2}$ min^{-1} of the acid-labile sulphate in sulphated chitosan indicated a reliable formation of the *N*-sulphate bond to the amino group in sulphated chitosan.

To obtain chitosan nitrate, 200 mg chitosan was stirred into a mixture of 10 ml acetic anhydride, 10 ml of acetic acid and 13 ml of absolute nitric acid at 0–5 °C for 5 hr.[10] After centrifugation the residue was washed with acetic acid and ether; the yield was 330 mg of nitrate ester of chitosan nitrate. The product corresponded to the formula:

$$C_6H_7O_2(ONO_2)_{1{\cdot}65}\,(OH)_{0{\cdot}35}\,(NHCOCH_3)_{0{\cdot}15}\,(NH_3^+NO_3^-).$$

Chitosan (85% free amine) was found to dissolve in absolute nitric acid but the resulting chitosan nitrate was nearly identical to that obtained employing absolute nitric acid admixed with acetic acid and acetic anhydride which react heterogeneously with chitosan. Both reaction media afforded the nitric acid salt of chitosan nitrate ester in which approximately 85% of the two available hydroxyl functions were esterified, corresponding to a degree of substitution of 1·7.

The 3,6-dimethyl ether of chitosan was studied as a reference compound in connection with the degradative studies of carboxyl reduced heparin.[24] The ether should be obtained from chitosan, as ethylation of chitin is quite difficult.[25] Thirty-eight grams of chitosan were dissolved with stirring in 1 l of 1 N hydrochloric acid. Sodium hydroxide pellets were dropped into the solution, 500 g in 1 l., and a thick paste resulted to which 500 ml water were added, always under stirring. Two hundred ml of dimethylsulphate were introduced during 1 hr with cooling and the mixture was left under stirring 8 hr, whereupon 40 g of sodium hydroxide and 40 g of dimethyl sulphate were carefully added. The latter operation was repeated twice more. The mixture was then neutralized with concentrated hydrochloric acid and dialysed; yield 36 g of a product corresponding to $C_6H_{10}O_2ClN(OCH_3)_2$.

Chitosan hydrochloride can be obtained by dissolving a gel of precipitated chitosan in 0·1 N hydrochloric acid at 50 °C, adding concentrated hydrochloric acid until the precipitation is complete and cooling to room temperature.[7] The salt can be dissolved in boiling water, from which it can be precipitated again with further concentrated hydrochloric acid. It should be remembered here that concentrated hydrochloric acid at high temperature can produce degradation of the polymer. Purified chitosan hydrochloride, 5 g, was dissolved in 100 ml of water, and placed in a polythene bottle containing 900 ml of concentrated hydrochloric acid at 53 °C. The sealed container was kept warm for 72 hr: this produces hydrolysis to the monosaccharide. Shorter periods give considerable quantities of higher oligosaccharides. The chitosan hydrolysate was analysed by gradient ion-exchange techniques on a sulphonic acid resin. The hydrolysis of chitosan with hydrochloric acid can be carried out in various conditions, for instance by 4 N hydrochloric acid at 100 °C.

Studies on glucosammonium chloride have been carried out in connection with the research on the reactions of polyols with boric acid. It was reported that glucosamine, unlike most polyols, does not form a complex with boric acid and borates to enhance the acidity of the acid. Nevertheless, some doubts have been raised on these conclusions.[(26-28)]

Epichlorohydrin derivatives of chitosan have been prepared and studied:[29-31] chitosan, benzylchitosan, and styrene oxide chitosan were treated with epichlorohydrin. Strongly basic polymers resulted with high stability and low degree of swelling in alkali and acids, and were proposed as chromatographic supports for the resolution of D- and L-mandelic acids. A way of preparation of the said derivative is as follows: 10 g of chitosan in mixture with 5·8 g of epichlorohydrin are boiled with 100 ml water for 1 hr, and the resulting product is heated at 120 °C with 70 ml of 4 N NaOH for 2 hr and the resulting derivative weighs 13·5 g. The epichlorohydrin acts as a cross-linking agent with the hydroxyl groups of the glucose units.

Molecular weight[32, 33]

The acid treatment of crab shells and the deacetylation procedures of Broussignac and Hackman do not lead to extended degradation of the polymer chain, as indicated by the average molecular weight obtained by light scattering on the polymer as obtained with no dialysis treatment.

Light scattering measurements were done at 25 °C with a Mod. 42000 Sofica photometer equipped with cylindrical cells in pure benzene. The instrument was set with dust-free benzene having a Rayleigh ratio $R_{90} = 16{\cdot}3 \times 10^{-6}$ at 546 nm.

The refractive index increments were measured at four distinct concentrations of chitosan in 8·5% formic acid + 0·5 M sodium formate and are presented in Fig. 4.1; from these the value $(dn/dc) = 0{\cdot}174$ ml g^{-1} was obtained. As far as chitosan solutions in formic acid are concerned, —NH_3^+ groups are responsible for the polyelectrolyte behaviour of the polymer, while the $HCOO^-$ counter ions are free to move.

At relatively high concentrations, the polyelectrolyte molecules overlap each other, and the counter ions are not allowed to leave the molecular domain. As dilution increases, and counter ions diffuse to regions where the polymer molecules are absent; therefore, the total electrical

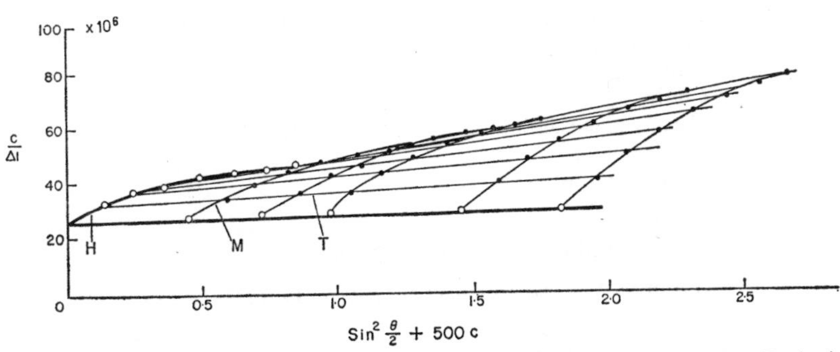

FIG. 4.4. Light-scattering measurements on chitosan in formic acid. (From R. A. A. Muzzarelli *et al.*, *Talanta* **19**, 1222 (1972).)

charge on each polymer chain increases and forces the molecule to extend itself to a larger configuration. For these reasons sodium formate was used in addition to the solvent, to depress the polyelectrolyte effect.

For each chitosan concentration, light scattering was measured, and the value C/I was plotted vs. $(\sin^2 \theta/2+500C)$ according to Zimm[34] in Fig. 4.4: the form of the diagram indicates a limited degree of polydispersion in agreement with data reported by other authors,[14] and with data on the chitosan precipitation.[4] The extrapolated average molecular weight value obtained by Muzzarelli and coworkers[32, 33] was 120,000. The same value was reported by Nagasawa and coworkers,[22, 23] who independently carried out their work about the same time.

Elemental analysis[32]

In Table 4.3 data is reported on the nitrogen content of chitin and of chitosan obtained in slightly different ways (different composition of the deacetylation mixtures) derived from the Broussignac preparation.

TABLE 4.3. ELEMENTAL ANALYSIS OF CHITOSAN SAMPLES, CHITIN AND DIETHYLAMINOETHYLCELLULOSE; *N*-ACETYLGLUCOSAMINE REFERENCE

Polymer	C	H	N
Chitosan (1st prep)	39·99	6·80	7·40
Chitosan (2nd prep)	39·66	6·22	7·40
Chitosan (3rd prep)	39·82	6·10	7·33
Chitin	43·53	6·12	6·26
N-acetylglucosamine	43·53	7·15	6·26
DE Cellulose			1·25*
PAB Cellulose			1·25*

* Certified by Whatman and Bio-Rad.

(From R. A. A. Muzzarelli *et al.*, *Talanta* **19,** 1222 (1972).)

In the elemental analysis measurements, chitosan ranks first, as the polymer with higher nitrogen content. The nitrogen per cent values are consistent on varying preparation conditions, which correspond to those selected for preparation of chitosan samples for molecular weight determinations.

One can easily establish a relationship between per centage nitrogen and per centage collection of transition metal ions from reference solutions. The trend stresses the fact that the nitrogen electrons play a prominent action in the metal ion fixation to the polymer; of course, other groups give their contribution too.

It is surprising that no author made an analysis for metals and other elements on chitin and chitosan. As these polymers are obtained from natural products, and a decalcification step is required, one would reasonably make a check for calcium and other elements. However, there is no mention in the literature of any analysis in this respect.

Several stocks of chitosan obtained following the Broussignac method, gave the data in Table 4.4.

TABLE 4.4. TRACE METAL CONTENT OF CHITOSAN (ppm)

Aluminium	5–50
Calcium	10–150
Copper	2*
Iron	10–100
Lead	<10
Manganese	< 1
Silicon	10–100
Titanium	< 1
Silver	< 1
Chromium	2·2*
Cobalt	0·2*
Zinc	0·3*

* By neutron activation analysis, otherwise by emission spectrography.

(From R. A. A. Muzzarelli *et al.*, *J. Chromatogr.* **47**, 414 (1970).)

Of these elements, aluminium, calcium, silicon and possibly iron are from the original material itself while chromium and copper come from the stainless steel reactor and from distilled water, respectively. These trace metals can be removed by a simple washing with EDTA 0·1 M and by rinsing with pure distilled water. Very little can be done about aluminium and silicon.

X-ray diffraction and i.r. spectrometry[32]

Diffraction spectra of chitin and chitosan show some structural resemblance between the two polymers.

TABLE 4.5. OBSERVED *d* VALUES (Å) FOR CHITOSAN AND METAL ION TREATED CHITOSAN (200 mg polymer in 50 ml of 0·044 M solution for 30 min)

Chitosan	Chitosan $+CuSO_4$	Chitosan $+CuCl_2$	Chitosan $+CdSO_4$	Chitosan $+CdCl_2$
8·49	6·99	5·65	9·01	9·42
4·46	6·37	5·41	5·07	5·11
3·02	5·32	2·74	4·36	4·40
2·48	4·44	2·55	3·01	2·45
2·27	3·90	2·38	2·47	2·28
2·06	3·18	2·24	2·27	2·09
1·90	3·01		2·17	1·91
1·86	2·91		1·90	1·87
	2·66		1·86	
	2·60			
	2·50			
	2·44			
	2·39			
	2·35			
	2·25			
	2·17			
	2·13			
	2·08			
	2·01			
	1·95			
	1·90			
	1·86			
	1·80			
	1·73			
	1·59			
	1·55			
	1·53			
	1·49			
	1·42			

(From R. A. A. Muzzarelli *et al.*, *Talanta* **19**, 1222 (1972).)

In Table 4.5, the spectrum relevant to copper sulphate treated chitosan shows very numerous and sharp bands of a new highly crystalline structure. It is a well-known fact that when a matrix substance crystallizes with a "solute" additive, one can observe either a matrix lattice parameter alteration, or the formation of a new crystalline phase which is characterized by new bands in the X-ray diffraction pattern, the latter case occurring when the "solute" atoms occupy well-defined positions in the matrix lattice.

Copper ions take definite and fixed positions in the chitosan lattice and, in view of the above-described nature of chitosan, one can therefore support the hypothesis that a true chemical bond is established between the chitosan nitrogen atoms and the copper ions. Anion contribution is not excluded, as it is known that chitosan sulphate is crystalline. Copper chloride treated chitosan yields less clear X-ray diffraction patterns.

The same trend is not observed in X-ray diffraction patterns recorded on chitosan treated with other metal ions: see for instance the cadmium treated chitosan X-ray diffraction pattern; however, they are expected to be retained by dative bonds in the light of the chitosan chelating ability so far proved with copper. Of course, copper exhibits a privileged characteristic, but the other ions are also introduced into the lattice ordinately and without destroying or altering the structure. Should a physical adsorption occur, a less ordered chitosan structure would be observed, but this is not the case.

After metal ion collection, in these cases chitosan structure is preserved in band number, intensity and sharpness.

Crystallinity is not lost during chromatography on chitosan powder. In fact X-ray diffraction on samples taken from the top of chromatographic columns, where copper was collected from 0·44 mM solutions and eluted with 0·01 M EDTA, shows that crystallinity becomes better as cycle number increases. This is accompanied by an increase in the amount of EDTA solution necessary to elute the same amount of copper.

Gamma radiation resistance of chitin and chitosan[35]

While studies[36, 37] on radiolysis of glucosamine and its monomer derivatives in aqueous solutions, conclude in favour of a high sensitivity of amino sugars to ^{60}Co radiations (leading to formation of mixtures of degradation products among which predominates 1,3-dihydroxyacetone) the corresponding polymers chitin and chitosan are rather indifferent to ^{60}Co radiations, provided that they are irradiated in form of solids.

The polymers were obtained from crab shells according to Broussignac.[12] Ultraviolet spectra on the washings of the final product, did not show any dissolved matter. Twenty milligrams were suspended in the solution in open vials for irradiation to simulate operative conditions for high radioactivity chromatography. Radiolysis in solution was verified by 0·5% solution of chitosan in 0·1 N perchloric acid.

From the data presented in Table 4.6 it can be noted that irradiation of chitosan does not involve a collection capacity reduction for most ions; remarkably, copper at pH 2·2 is collected better on irradiated chitosan than on plain chitosan.

TABLE 4.6. PERCENTAGE COLLECTION OF METAL IONS ON 200 mg CHITOSAN POWDER, 100–200 MESH, FROM 50 ml OF 0·5 mM SOLUTIONS. CHITOSAN WAS IRRADIATED 1 YR BEFORE USE

Metal	pH	Non-irradiated	Irradiated
Iron	4·5	79	80
Copper	5·0	86	89
	2·2	2	50
Zinc	5·4	71	67
	2·3	12	20
Arsenic	6·0	25	0
	2·5	100	15
Silver	2·5	80	94
Tin	2·5	17	0
Mercury	5·8	96	97
	2·3	100	100

(From R. A. A. Muzzarelli *et al.*, *J. Radioanal. Chem.* **12,** 431 (1972).)

The results in Fig. 4.5 show that chitosan suspended in acids at various pH values can tolerate gamma irradiation of about 50,000 krad, without losing its identity. In fact the i.r. spectra reported, correspond to the i.r. spectrum of the pure untreated chitosan. In the solution at pH = 2·5 a very small fraction of chitosan might be present in dissolved form, and chitosan, once dissolved, can undergo radiolysis. In fact, the u.v. spectra of the filtered solution at pH = 2·5 from the irradiation experiments, show that traces of chitosan were radiolysed. These spectra correspond to those taken on perchloric acid solutions of chitosan. In any case, this

TABLE 4.7. PER CENT COLLECTION OF SEVERAL METAL IONS ON 7×1 cm COLUMNS OF CHITOSAN 100–200 MESH FROM 50 OR 100 ml OF SOAP SOLUTION. pH = 4·5. DATA IS FOR BOTH IRRADIATED AND NON-IRRADIATED CHITOSAN

Metal added	Per cent collection from 50 ml	from 100 ml
^{95}Zr	87	68
^{125}Sb	82	75
^{106}Ru (III)	60	45
$^{106}Ru\ (NO)^{2+}$	5	5
^{60}Co	100	54
^{144}Ce	100	85

The per cents are calculated on the basis of radioactivity by respect to the amount contained in 50 or in 100 ml respectively.

(From R. A. A. Muzzarelli *et al.*, *J. Radioanal. Chem.* **12**, 431 (1972).)

small amount of chitosan being lost and destroyed is of no significance in column chromatography involving nitric acid solutions at pH = 2·5. Suspensions of chitosan powder in hydrochloric acid can equally tolerate irradiation without alteration of the i.r. spectrum taken on the powder. Also chitin and chitosan polymolybdate are not decomposed under irradiation.

The measurements done under these conditions clearly indicate that chitosan powder is far from being destroyed by gamma irradiation and therefore can be used in chromatography of high specific activity radioactive products. In fact, while chitosan exhibits a good collection ability in concentrated soap, chitin is not able to collect metal ions from the soap solution. For chitosan, Table 4.7 presents the percentage retention on the column for the first 50 ml and for the total 100 ml amount of

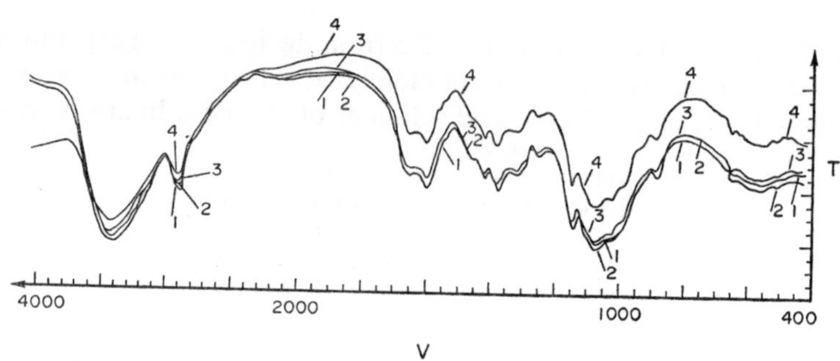

FIG. 4.5. Infrared spectra of chitosan powder suspended in hydrochloric acid at pH = 5·1 and irradiated with 23,500 krad (1); the same in water (2); the same as dry powder (3), compared with the spectrum of non-irradiated chitosan (4). (From R. A. A. Muzzarelli *et al.*, *J. Radioanal. Chem.*, **12**, 431 (1972).)

soap solution. These data refer to breakthrough experiments. For the elements listed there is no appreciable difference between collection on irradiated or non-irradiated chitosan powders.

For Sb and Ru the loss from the column did not follow a breakthrough curve, while Zr and Co showed a typical breakthrough curve. The column however, was not yet saturated when the first leakage occurred, due to various chemical species of cobalt produced by the soap components.

As molecules similar or related to chitosan were studied by pulse radiolysis, some results are recalled here. Heparin, hyaluronic acid and carboxymethylcellulose with other polymers and their reactions with e^-_{aq} and hydroxyl radicals have been studied by pulse radiolysis techniques.[38] Comparison of the hydroxyl rate constants of the polyanions with those of the related monomeric repeating units indicated that the nature of the polyanion in solution exerts an important influence on reactivity. The reactions of e^-_{aq} with the polyanions are generally slow except for hyaluronic acid.

In fact, in irradiations of aqueous solutions, hyaluronic acid exhibits a significant reactivity:[39] for the reaction e^-_{aq}+D-glucose 5×10^{-2} M the rate constant is 10^6 M^{-1} sec^{-1}, while for hyaluronic acid the constant is 10^9 M^{-1} sec^{-1}. Extreme excitation following gamma-irradiation can remove the oxygen *p*-electron to produce the intermediate II that would undergo glycosidic cleavage, which is a demonstrated consequence of radiation action. The intermediate carbonium ion in III would react rapidly with water, and the presence of the uronic acid group can labilize the hydrogen atom at C-5. The predominating singlet for irradiated hyaluronic acid at high doses and guluronic acid points to elimination of the H at C-5 and the radical would be stabilized by resonance interactions with the carboxylic group

Following an original method for investigating the ion-binding between polyanions and their associated counterions the interactions of the cationic dye methylene blue and the anionic sites of heparin were

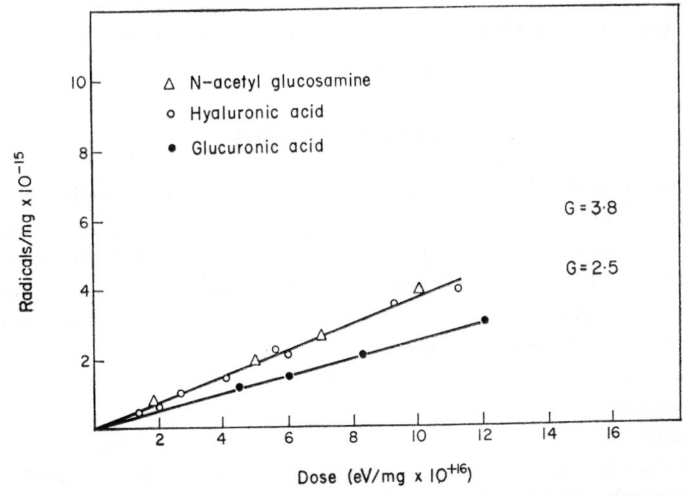

FIG. 4.6. Radical yields from gamma-irradiated *N*-acetylglucosamine, hyaluronic acid, and glucuronic acid. (From E. A. Balázs *et al.*, *Rad. Res.* **31**, 243 (1967).)

studied.[40] Ion-binding by short-range interaction forces, leads to charge neutralization of the counter-ion with subsequent unreactivity towards e^-_{aq}. Changes in the e^-_{aq} reaction rate with methylene blue and acridine orange, following the addition of polyanions demonstrated that interactions occur in aqueous solutions between the dyes and hyaluronic acid, heparin and DNA as well as other polymers. The method was extended to study the interactions between the polycations polylysine, and protamine sulphate, and the polyanions heparine and DNA before electrochemical equivalence and consequent precipitation.[41] Ion-binding due to the formation of contact-ion pairs, formed usually in water as a result of the high electrostatic potential on the surface of the macro-ion accounted for the results.

Complexes of other dyes like thiazin, acridine and triphenylmethane with glycosaminoglycans were reported.[42–44] Heparin and hyaluronic acid formed complexes exhibiting light-induced paramagnetism.

In any case, while the reported studies refer to the polymers in general terms like aminopolysaccharides and glycosaminoglycans, chitin and chitosan have not yet been studied by pulse radiolysis. For *N*-acetylglucosamine only radical yields have been published, in Fig. 4.6.

AGGREGATIONS AND COAGULATIONS OBTAINED WITH CHITOSAN

Biomedical research

Chitosan is rarely found as such in nature. *Chlorella* cells probably contain chitosan in their cell walls.[45] By fractionating the cell material, it was revealed that most of glucosamine was present in the material insoluble in ethanol, ethyl ether, and aqueous perchloric acid. Chitosan is sometimes reported as a component of secretions, like the spermatophore stalk of *Podura aquatica*.[46]

At a concentration of 2×10^{-5} M (on the basis of glucosamine content) chitosan completely inhibits the fermentation of glucose by a 0·03 per cent yeast suspension:[47] this corresponds to an effectiveness ten times higher than for the detergent dodecyltrimethylammonium bromide. For this research chitosan was dissolved in acetic acid and dialysed against distilled water before use. Under most conditions, it appeared that the inhibition is practically irreversible. Under the same conditions chitosan inhibits the fermentation of 0·03 M glucose 6-phosphate, fructose and sucrose. The following cations other than chitosan were found to inhibit the fermentation of glucose by yeast: diethylaminoethylcellulose, protamine sulphate, uranyl acetate, aluminium sulphate, aniline and -D glucosamine, but none was as effective as chitosan at equivalent concentration. Short-chain amines did not inhibit fermentation.

Chitosan at the said concentration caused an aggregation of yeast cells, observable at the microscope. At higher chitosan concentration, 2×10^{-3} M, the yeast was not aggregated. It was suggested that chitosan is electrostatically bound to yeast, and the same happens with diethylaminoethylcellulose. Experimentally, it was demonstrated that chitosan acts on the outer cell wall, and experiments with chitosan hydrolysates showed that the degree of inhibition was proportional to the length of the polymer chain below concentrations of about 4×10^{-5} M, but above that, yeast was fully inhibited by short chain polyglucosamines (14 and 74 glucosamine units).

It would appear that chitosan exerts its inhibiting action by forming a layer on the yeast cell surface which prevents the entry of glucose, or other substances into the cell. Interaction between chitosan amino groups and anionic groups on the yeast cell surface takes place.

Chitosan ranks first among the substances that enhance the β-glucuronidase activity; protamine, serum albumine, DNA, 1,10-diaminodecane, gelatin, lysine and ornithine, to name a few are not such good activators as chitosan, as the data shows in Table 4.8. The experimental findings support the view that this series of activators promote the recombination of the dissociation products to reconstitute the active enzyme.(48)

In the physico-chemical study of the interactions of macromolecular polyanions with human serum β-lipoproteins,(49) the complexes resulting from the interaction of chitosan sulphate were the only ones defined as soluble complexes with little or no tendency to dissociate (Type II) while insoluble complexes (Type I) were formed by corn amylopectin sulphate, and the other polyanions formed soluble complexes which readily dissociate into their components, as summarized in Table 4.9. The presence of hexosamine units in the polyanion, either *N*-acetylated or *N*-sulphated, reduces markedly the affinity between the β-lipoprotein and the polyanion. It was remarked that a certain analogy between the solubility of the barium salts of the polysaccharide sulphates and that of

TABLE 4.8. POTENCY OF β-GLUCURONIDASE ACTIVATORS

Activator		Concn. of half-activation γ/ml
Chitosan		2
Protamine		3–4
Cryst, bovine serum albumin		3–4
Desoxyribonucleic acid (Krishell)		8·5
Gelatin		14
Cryst. chymotrypsin		16
Yeast ribonucleic acid		300
$H_2N(CH_2)_n\ NH_3$	$n = 10$	7
	$n = 8$	60–90
	$n = 7$	200
	$n = 6$	1300
(Cadaverine)	$n = 5$	250
(Putrescine)	$n = 4$	1000
	$n = 3$	300
	$n = 2$	3000
Spermine		60–80
Spermidine		300
L-Lysine		3000
DL-Ornithine		13000
L-Arginine		18000

(From P. Bernfeld *et al.*, *J. Am. Chem. Soc.* **76,** 4872 (1954).)

their complexes with β-lipoproteins exists at pH = 8·7. It was also assumed that the rather high solubility of the barium salts of hexosamine-containing polysaccharide sulphates is due to chelate formation and therefore, it is expected that hydrogen bonding takes place in the complexes of these polyanions with β-lipoproteins.

Chitosan, at a concentration ranging from 0·05 to $0{\cdot}006 \times 10^{-3}$ meq ml^{-1}, selectively aggregates the leukaemia cells *in vitro* when compared to erythrocytes and bone marrow cells.[50] On the contrary, poly-L-lysine and polyethylenimine at these concentrations do not discriminate between the erythrocytes and the leukaemia cells. Chitosan produced dense aggregations, cytoplasmatic swelling and/or cytolysis of the leu-

TABLE 4.9. INFLUENCE OF PRESENCE OF HEXOSAMINE IN SULPHATED POLYSACCHARIDES ON THEIR INTERACTION WITH β-LIPOPROTEIN

Polyanion	No. of sulphate groups per repeating unit	Nephelometric index (relative values)	Type of interaction from electrophoretic analyses			
			Type I	Type II	Type III	None
Corn amylopectin sulphate	2·18	100*	+++			
Sulphated hyaluronic acid†	0·58–1·33**	1			+++	
Sulphated hyaluronic acid†	0·24§	1			+	++
Sulphated hyaluronic acid‡	0·66§	1			+++	
Hyaluronic acid†‡	0	0				+++
Sulphated deacetylated hyaluronic acid†	2·30§	1			+++	
Chondroitin sulphate A	0·94§	0				+++
Sulphated chondroitin sulphate A	1·6–2·35§	1			+++	
Sulphated deacetylated chondroitin sulphate A	2·03§	1			+++	
Chondroitin sulphate C	1§	1				+++
Chondroitin sulphate B	0·77§	1			+++	
Heparin**	1·45–2·68**	0			+++	
Sulphated chitin	0·68	1			+++	
Sulphated deacetylated chitin (chitosan)	0·23	1		++	+	

* Reference polyanion.
† From bovine vitreous body.
** Sulphate per disaccharide unit.
‡ From human umbilical cord.
§ General Biochemicals Inc. and Liquaemin Sodium, Organon.

(From P. Bernfeld *et al.*, *J. Biol. Chem.* **235,** 2852 (1960).)

kaemia cells and increased the permeability of these cells to trypan blue. Poly-L-lysine produced similar morphological alterations of the erythrocytes while chitosan does not alter their morphology. Chronic treatment of mice with chitosan, after intraperitoneal implants of ascites tumour, retarded the dissemination of the leukaemic cells into the peripheral blood. The chitosan stock used for these experiments was a partially deacetylated chitin containing equal portions of *N*-acetylglucosamine and free glucosamine. Its molecular weight was 60,000–140,000.

The cytoplasmatic swelling effect and the haemolysis or cytolysis manifested by polycation-treated cells suspended in an isotonic saline medium suggest that the adsorption of cationic polyelectrolytes by the cell surface precedes a disruption of the structural and conformational integrity of the plasmalemma, which alters the cell permeability to solutes. This introduces changes in the physiological concentration gradient existing between the cells and the saline suspension medium and thus induces, through osmotic swelling, a mechanical rupture of the plasmalemma.

An unique feature of cancer cell so far discovered is the physicochemical property of its surface.[51] The aggregation of normal and tumour cells induced by a polycation offers a means of exploring further the uniqueness of the tumour cells surfaces. Possibly, a polycation can be found with the desired degree of specificity for tumour cell-surface adsorption; for instance, in the opinion of the author, completely deacetylated chitosan and chitosan derivatives should be tested. Using polycations chemically modified, more information on tumour cell surface characteristics may be reached, and novel antitumour agents acting primarily on the cell surface may be developed. As far as we know polycations can kill tumour cells *in vitro* and can alter the degree of tumour growth *in vivo*.

Chitosan is not toxic to mice:[52] a daily intake of 18 g kg^{-1} body weight for a 19 day period resulted in a slight toxicity. Sixteen grams of chitosan formate and 14 g of chitosan acetate, under the same conditions, showed similar results, with a somewhat greater toxicity for the acetate.

Chitosan shows a moderate but remarkable anti-ulcer and anti-acid properties.[53] The anti-ulcer activity is due to its capacity to bind free gastric acid and to a significant ability to act as demulcent. Gastric demulcent activity of chitosan was evident in dogs given large doses of aspirin. Anti-peptic activity was also observed with chitosan coprecipitated with aluminium hydroxide: this preparate was almost equal to carrageenin in inhibiting proteolitic activity of gastric secretions.

Technological applications

A use for partially deacetylated chitin (a chitosan with less than 7·0% nitrogen content) was provided in that it may be used in treating turbid matter in an aqueous solution to help remove the turbid matter.[54] Chitosan is used as a viscosity builder to settle the solids suspended in the liquid. Chitosan forms agglomerates comprising the impurities and chitosan itself, and it is presented as a coagulant. The impurities considered were insoluble substances, mainly compounds of alkali earth elements. In practice the authors presented a physical process, where the long molecules of dissolved chitosan wrap the solid particles of the turbid matter suspended in the liquid, and bring them together to form an agglomerate. The resulting agglomerate has to be discarded in all cases. In several cases the action of chitosan can be compared to that of aluminium or ferric salts. For this process there is interest in keeping the amount of dissolved chitosan used, as low as possible.

For example, a suspension of montmorillonite was prepared by adding 10 g of sieved clay to 1 l of distilled water and allowing the suspension to settle for 24 hr. The turbid supernatant liquid was found to contain 224 ppm of suspended solids and the Jackson turbidimeter reading was 350. The results of the various treatments on this solution are reported in Table 4.10. It can be seen that the use of chitosan in conjunction with both alum and lime may reduce the turbidity to less than one. Chitosan is as effective as Separan, a commercial flocculating agent. Similar results were obtained with kaolinite clay suspensions. A variety of chitosan obtained from partially deaminated chitin submitted to partial deacetylation, was also used to study the effect of nitrogen content and acetyl content on the coagulating ability of the polymer. Data in Table 4.11 show that high nitrogen content, low deacetylation samples, give the best results, provided they are used in small concentrations: in fact there is interest in keeping as low as possible the amount of polymer; there is more uniformity in the results for montmorillonite. Several applications were foreseen: treatment of waste waters from miming operations; treatment of hard water or temporary hard water. Also removal of

TABLE 4.10. SETTLING OF TURBID MATTER (MONTMORILLONITE AND KAOLINITE) BY CHITOSAN OR SEPARAN (VALUES ARE EXPRESSED IN ppm)

	Montmorillonite samples						Kaolinite clay samples					
Alum	100	100	100	100	100	100	100	100	100	100	100	100
Lime		2·5	12·5		2·5	12·8		25	12·5		25	12·5
pH	4·3	6·4	4·7	4·3	6·6	4·0	4·0	7·0	4·5	4·3	7·1	4·7
Chitosan	1	1	1			1	1	1	1			
Separan				1	1					1	1	1
Turbidity after												
30 min	4	<1	<1	4	<1	<1	165	<1	165	32	<1	205
60 min	4	<1	<1	4	<1	<1	152	<1	132	27	<1	165

(From Q. P. Peniston *et al.*, U.S. Patent 3,533,940 (1970).)

TABLE 4.11. SETTLING OF TURBID MATTER (MONTMORILLONITE AND KAOLINITE) BY CHITOSANS OF VARIOUS DEGREES OF DEAMINATION AND DEACETYLATION

	Montmorillonite samples						Kaolinite clay samples					
Nitrogen %	5·5	5·5	7·8	7·8	7·6	7·6	5·5	5·5	7·8	7·8	7·6	7·6
Deacetyl. %	?	?	53	53	39	39	?	?	53	53	39	39
Chitosan ppm used	1	0·5	1	0·5	1	0·5	1	0·5	1	0·5	1	0·5
Turbidity after minutes												
30	18	32	20	36	28	48	~270	140	~300	125	~225	40
60	18	26	11	27	25	27	~270	115	~300	95	~240	38
120	18	24	11	25	21	25	~270	90	~300	80	~240	30
180	12	22	8	22	15	22	~240	75	~240	67	~190	30

Turbidimeter readings for montmorillonite before treatment: 62–40;
for kaolinite before treatment 370–350.
(From Q. P. Peniston *et al.*, U.S. Patent 3,533,940 (1970).)

tannin and polyphenolic materials from aqueous media can be realized with the use of chitosan.

Chitosan is therefore conceived as a clarification aid and viscosity builder in solutions for rapid settling of suspended solids. In the said publication[54] there is no mention of possible interactions with transition metal ions or compounds and the conditions reported in the examples and in the text would not allow recovery of metal ions.

As the adsorption of lignosulphonates present in the spent liquors from the sulphite pulping of wood can be performed with chitin,[55] it could be anticipated that the application of the interpolymeric adsorption technique to commercial agar would lead to the preferential retention of the highly sulphated impurities associated with commercial agarose. This in fact proved to be the case, and it was shown that commercial agar can be fractionated by using chitin or chitosan.[56] Two experimental systems were investigated: a homogeneous procedure in which a 2% solution of agar in formamide mixed 2 : 1 v/v with a 0·5% solution of chitosan in 3% acetic acid, pH = 4·5; the insoluble agaropectin (the sulphated polymer) precipitated with chitosan from the agarose-rich supernatant liquor. The heterogeneous procedure was carried out on a 2% solution of agar in formamide, the solution was stirred for 4 hr with an equal weight of chitosan or chitin. After centrifugation the clear solution was treated with ethanol to collect agarose. The sulphur content of the separated compounds was determined by neutron activation analysis and gamma-ray spectrometry.

Other anionic polymers could be purified by a similar procedure, like alginic acid, the sulphated extracts of *Ulva lactuca* and cellulose sulphate, which are precipitated by chitosan.

CHITOSAN MEMBRANES

Owing to its solubility in formic or acetic acid, chitosan is apt for the preparation of a membrane: therefore a chelating membrane becomes feasible on the basis of the film-forming ability and of the chelating properties of chitosan.

There is interest in making chelating membranes available for preconcentration of trace elements, for selective isolation of certain elements from saline solutions, or for special sample preparation in nuclear chemistry or in electrochemistry. A chelating membrane is expected to exhibit superior characteristics as it would be selective, would not release ions to solution, and would bind transition metals by dative bonds instead of ionic bonds.

The chitosan membranes were described and characterized for the first time by Muzzarelli and coworkers.[57]

In addition to a simple technique based on the evaporation of formic acid from a chitosan solution spread on a glass plate, there is another approach to the preparation of a chitosan membrane. A chitinous membrane was reported to occur in the shell of the cuttlefish *Sepia officinalis;* to isolate the chitin membrane the shell of the cuttlefish was first decalcified and the resulting chitin+protein complex was deproteinized (see Fig. 3.1 and relevant text).

As soon as a chitin membrane is available, it can be deacetylated to yield a chitosan membrane. The form and area of the membrane so obtained are those of the shell.

Chitosan membranes were therefore obtained in both ways: artificial chitosan membranes were produced from a solution of 1 g of chitosan (m.w. 120,000) in 100 ml of 65% reagent-grade formic acid. This solution was poured on a 20×20 cm glass plate kept perfectly horizontal upon a water bath. The solution was prevented from dripping out of the plate by a 1 mm board of Silastic. When formic acid was completely evaporated, the glass plate was exposed to cold air so that the membrane could be easily taken away from the glass surface. This water-soluble chitosan formate membrane was dipped into 1 N NaOH and rubbed vigorously with gloves, at the same time avoiding folding the membrane. Thereafter the membrane was washed with distilled water to neutrality, and then was laid down on a clean glass plate and kept at 60 °C until dry.

To produce a chitosan membrane from the *Sepia officinalis* shell, the isolation procedure included a 10% hydrochloric acid treatment for 3 or 4 days, as previously reported, an ether extraction and a deacetylation step according to Broussignac. The chitosan identity was checked by infrared spectrophotometry. Both procedures involving chitosan powder from crab shells, or the *Sepia officinalis* membrane, yielded membranes perfectly transparent and of high mechanical resistance.

In fact, the tensile strength was found to be 7 kg mm^{-2} in both directions. No appreciable elongation was observed. These data are in agreement with data incidentally reported by other authors in a paper on the preparation of chitosan.[14] The burst strength was found to be 1·8 kg cm^{-2}.

The permeabilities to gases and vapour are reported in Table 4.12 and compared to data for a cellophane membrane. The outstanding

TABLE 4.12. PERMEABILITIES OF CHITOSAN MEMBRANES, AS MEASURED WITH A HONEYWELL WATER VAPOUR TRANSMISSION TESTER, AND WITH A VASCHETTI & GROSSO WITH DOW CELL 6129/C APPARATUS, FOLLOWING THE ASTM D 1434

Membrane	Thickness, μm	$g\ m^{-2}/24$ hr	Permeabilities $ml\ m^{-2}/24$ hr at 21 °C 760 mm		
		H_2O D = 90% at 100 °F	O_2	N_2	CO_2
Chitosan	20	1200	3·6	0·7	2·7
Cellophane 30 atg	22	1200	11·3	3·7	264·0

(R. A. A. Muzzarelli, original results.)

ability of the chitosan membrane to prevent the passage of oxygen, nitrogen and carbon dioxide is evident; its permeability to water moisture is comparable to that for cellophane.

While man-made chitosan membranes are amorphous in both salt and free amine forms, the chitosan membranes obtained from the *Sepia officinalis* show X-ray diffraction lines. As these lines correspond to those of the chitosan powder, as already reported,[32] the above finding indicates that a certain ordered structure can be expected in duly treated man-made membranes. In fact, it was observed at the microscope that alcohols make the membrane to crystallize upon wetting.

Also, crystallization occurs upon collection of transition metal ions (see Table 4.13), which is accompanied by colour appearance, e.g. yellow with Cr(VI), green with Cr(III) and blue with Cu. In each case particular crystallization patterns can be observed.

Chitosan membranes generally show lower capacity than chitosan powder, due to the reduced contact surface; however, collection is good, especially for molybdenum, chromate and mercury. The metal ions, once collected on the membrane, do not undergo the usual reactions currently used for their identification: for instance, chromate does not react with diphenylcarbazide, and the membrane keeps its yellow colour, while the solution remains colourless.

Upon stirring a 20 mg piece of membrane 20 μm thick for 10 minutes in 50 ml solution, it is possible to estimate the chromate concentration

TABLE 4.13. PER CENT COLLECTION OF TRANSITION METAL IONS ON CHITOSAN MEMBRANES (20 mg MEMBRANE IN 50 ml OF 1 ppm AQUEOUS SOLUTION, pH = 4·5, 12 hr STIRRING AT 20 °C, MEMBRANE THICKNESS 20 μm)

Chitosan membrane	Cr(III)	Cr(VI)	Mn	Fe(III)	Co	Ni	Cu	Hg	Zn	Mo	As	Ag	Pb
Man-made	37	92	0	—	0	0	42	90	15	100	0	69	0
Man-made (ethanol treated)	4	75	0	6	0	0	52	82	15	85	0	87	51
Sepia officinalis	60	20	0	80	0	10	50	85	15	90	0	60	60

(R. A. A. Muzzarelli, original results.)

down to 0·3 ppm by visual comparison with similarly treated membranes in reference solutions. Copper and molybdenum can be estimated down to 1·0 ppm.

As for electrical conductivity, the chitosan membrane exhibits characteristics similar to those of the organic molecular crystals.

The physical and chemical properties of these membranes make them interesting for applications in many fields.

REFERENCES

1. F. HOPPE-SEYLER, *Ber.* **27**, 3329 (1894).
2. T. ARAKI, *Z. Physiol. Chem.* **20**, 498 (1895).
3. O. VON FURTH and M. RUSSO, *Chem. Zentr.* **77**, 133 (1906).
4. G. W. RIGBY, U.S. Patent 2,040,879 (1936).
5. G. L. CLARK and A. F. SMITH, *J. Phys. Chem.* **40**, 863 (1936).
6. S. E. DARMON and K. M. RUDALL, *Disc. Faraday Soc.* **9**, 251 (1950).
7. S. T. HOROWITZ, S. ROSEMAN and H. J. BLUMENTHAL, *J. Am. Chem. Soc.* **79**, 5046 (1957).
8. D. HORTON and D. R. LINEBACK, *Methods Carbohydr. Chem.* **5**, 403 (1965).
9. R. W. JEANLOZ and E. FORCHIELLI, *Helv. Chim. Acta* **33**, 1690 (1950).
10. M. L. WOLFROM, G. G. MAHER and A. CHANEY, *J. Org. Chem.* **23**, 1990 (1958).
11. M. L. WOLFROM and T. S. SHEN HAN, *J. Am. Chem. Soc.* **81**, 1764 (1959).
12. P. BROUSSIGNAC, *Chim. Ind. Génie Chim.* **99**, 1241 (1968).

13. T. Fujita, Japan Patent 7,013,599 (1970).
14. L. A. Nud'ga, E. A. Plisko and S. N. Danilov, *Zh. Obshchei Khimii* **41**, 2555 (1971).
15. R. G. Schweiger, *Chem. Ind.* 296 (1969).
16. G. G. Allan, P. G. Johnson, Y. Z. Lai and K. V. Sarkanen, *Chem. Ind.* 127 (1971).
17. M. L. Wolfrom, D. I. Weisblat, J. V. Karabinos, W. H. McNeely and J. McLean, *J. Am. Chem. Soc.* **65**, 2077 (1943).
18. J. E. Jorpes, H. Bostrom and V. Mutt, *J. biol. Chem.* **183**, 607 (1950).
19. M. L. Wolfrom and W. H. McNeely, *J. Am. Chem. Soc.* **67**, 748 (1945).
20. A. B. Foster, E. F. Martlew and M. Stacey, *Chem. Ind.* 825 (1953).
21. K. Nagasawa and N. Tanoura, *Chem. Pharm. Bull.* **20**, 157 (1972).
22. K. Nagasawa and Y. Inoue, *Chem. Pharm. Bull.* **19**, 2617 (1971).
23. K. Nagasawa, Y. Tohira, Y. Inoue and N. Tanoura, *Carbohydr. Res.* **18**, 95 (1971).
24. M. L. Wolfrom, J. R. Vercellotti and D. Horton, *J. Org. Chem.* **29**, 548 (1964).
25. P. Schorigin and N. N. Makarowa-Semljanskaya, *Ber.* **68**, 969 (1935).
26. H. K. Zimmerman, *Arch. Bioch. Biophys.* **75**, 520 (1958).
27. J. Boeseken, *Adv. Carbohydr. Chem.* **4**, 189 (1949).
28. P. W. Kent and M. W. Whitehouse, *The Biochemistry of Amino Sugars*, p. 166, Academic Press, London, 1955.
29. J. Noguchi and K. Arato, *Kogyo Kagaku Zasshi* **72**, 796 (1969).
30. J. Noguchi, Japan Patent 24,400 (1965).
31. J. Noguchi, S. Tokura, M. Inomata and C. Asano, *Kogyo Kagaku Zasshi* **68**, 904 (1965).
32. R. A. A. Muzzarelli, M. Pizzoli and A. Ferrero, *Talanta* **19**, 1222 (1972).
33. C. Guarnieri, D. Sc. Thesis submitted to the Faculty of Sciences of the University of Bologna, (referee R. A. A. Muzzarelli), 1969–70.
34. B. H. Zimm, *J. Chem. Phys.* **16**, 1093 (1948).
35. R. A. A. Muzzarelli and O. Tubertini, *J. Radioanal. Chem.* **12**, 431 (1972).
36. N. K. Kochetkov, L. I. Kudryashov and T. N. Senchenkova, *Dokl. Akad. Nauk. SSSR* **154**, 642 (1964).
37. N. K. Kochetkov, L. I. Kudryashov, T. N. Senchenkova and L. I. Nedoborova, *Zh. Obshchei Khimii* **36**, 1020 (1966).
38. E. A. Balazs, J. V. Davies, G. O. Phillips and D. S. Schenfele, *J. Chem. Soc.* **C**, 1420 (1968).
39. E. A. Balazs, J. V. Davies, G. O. Phillips and M. D. Young, *Rad. Res.* **31**, 243 (1967).
40. E. A. Balazs, J. V. Davies, G. O. Phillips and D. S. Schenfele, *J. Chem. Soc.* **C**, 1424 (1968).
41. E. A. Balazs, J. V. Davies, G. O. Phillips and D. S. Schenfele, *J. Chem. Soc.* **C**, 1429 (1968).
42. E. A. Balazs, G. O. Phillips and D. S. Schenfele, *Biochim. Biophys. Acta* **141**, 382 (1967).
43. E. A. Balazs, J. V. Davies, G. O. Phillips and D. S. Schenfele, *Biochem. Biophys. Res. Comm.* **30**, 386 (1968).

44. J. S. Moore, G. O. Phillips, K. S. Dodgson and J. V. Davies, *Proc. Biochem. Soc.* 18-P (1968).
45. S. Mihara, *Plant Cell Physiol.* **2,** 25 (1961).
46. W. Schliwa, *Zool. Jahrb. Abt. Anat. Ontog.* **82,** 445 (1965).
47. G. B. Ralston, M. V. Tracey and P. M. Wrench, *Biochem. Biophys. Acta* **93,** 655 (1964).
48. P. Bernfeld, H. C. Bernfeld, J. S. Nisselbaum and W. H. Fishman, *J. Am. Chem. Soc.* **76,** 4872 (1954).
49. P. Bernfeld, J. S. Nisselbaum, B. J. Berkeley and R. W. Hanson, *J. Biol. Chem.* **235,** 2852 (1960).
50. A. E. Sirica and R. J. Woodman, *J. Nat. Cancer Inst.* **47,** 378 (1971).
51. D. F. Wallach, *Proc. Nat. Sci. USA* **61,** 868 (1968).
52. K. Arai, T. Kinumaki, T. Fujita, *Tokaiku Suisan Kenkyusho Kenkyu Hokoku* 89 (1968).
53. I. W. Hillyard, J. Doczi and P. Kiernan, *Proc. Soc. Expl. Biol. Med.* **115,** 1108 (1964).
54. Q. P. Peniston and E. L. Johnson, U.S. Patent 3,533,940 (1970).
55. F. E. Brauns, U.S. Patent 3,297,676 (1967).
56. G. G. Allan, P. G. Johnson, Y. Z. Lai and K. V. Sarkanen, *Carbohyd. Res.* **17,** 234 (1971).
57. R. A. A. Muzzarelli, A. Isolati and A. Ferrero (submitted for publication).

CHAPTER 5

ANALYTICAL APPLICATIONS OF CHITIN AND CHITOSAN

INDIFFERENCE OF CHITIN AND CHITOSAN TO GROUP IA AND IIA IONS

When chitin or chitosan powders are brought in contact with alkali or alkali earth ions, no swelling, contraction, shrinking or colour appearance can be observed. Therefore, high concentration solutions were passed through small columns of chitosan made with about 2 g of 100–200 mesh powder. The columns were then washed with water and the ion concentration determined on 20 ml fractions taken progressively up to 700 ml water. This amount of water was found not to be detrimental for transition element retention on similar columns. The polymer was then taken out from the calibrated glass tube and burned at 700 °C. On the residue, dissolved in hydrochloric acid, the element was determined again. The results obtained are in Table 5.1. Low concentrations of Rb, Cs, and Ba were used, while Sr was used as carrier-free radioisotope only. Figures 5.1 and 5.2 illustrate the washing curves for sodium and magnesium chlorides and for potassium and calcium chlorides respectively.[1-3]

It was seen that the alkali and alkali earth element content of the polymer, after washing, is very low, especially compared to the relatively enormous amount of salt passed. For strontium which was used only as carrier-free, the per cent retained is relatively high, but it is possibly due to exchange with some calcium originally present in the polymer. However, it seems that strontium has higher affinity for these polymers than

TABLE 5.1. INDIFFERENCE OF 2 g OF CHITOSAN IN A 10×1 cm COLUMN TO ALKALI AND ALKALI–EARTH IONS. ATOMIC ABSORPTION AND EMISSION SPECTROPHOTOMETRY

Solution volume (ml)	content (g)	pH	Amount of metal ion on the polymer after washing with 700 ml water (mg)	Metal ion content of the last fraction (ppm)
Lithium chloride				
100	0·0032	6·2	absent	absent
Sodium chloride				
60	15·00	5·5	0·2	0·8
Sodium nitrate				
60	15·00	4·5	0·3	0·3
Potassium chloride				
100	24·00	6·0	0·02	1·0
Potassium nitrate				
100	10·00	5·5	0·02	0·2
Caesium chloride				
100	0·009	6·0	0·5	absent
Thallium nitrate				
100	1·00	6·0	20·0	absent
Magnesium chloride				
60	15·00	4·5	2·6	4·0
Magnesium nitrate				
60	50·00	7·0	1·0	3·0
Calcium chloride				
100	35·40	6·0	2·7	3·2
Calcium nitrate tetrahydrate				
100	98·73	6·2	1·2	4·5
Barium chloride				
100	0·021	5·5	0·0004	absent

(R. A. A. Muzzarelli, original results.)

other elements. The use of solutions at lower pH prevents these elements from remaining in the column even to such a low extent. The behaviour of caesium and rubidium, presented in Fig. 5.3, confirms that these metals are not collected by chitin and chitosan; the data in acidic solutions indicate that no collection at all takes place (see Table 5.2).

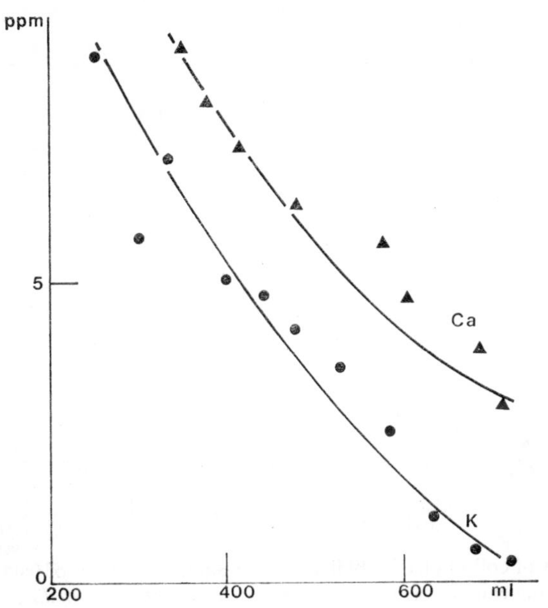

FIG. 5.1. Washing curves (metal ion concentration in ppm vs. water ml) recorded after passing a salt solution through a chitosan column (conditions listed in Table 5.1). (R. A. A. Muzzarelli, original results.)

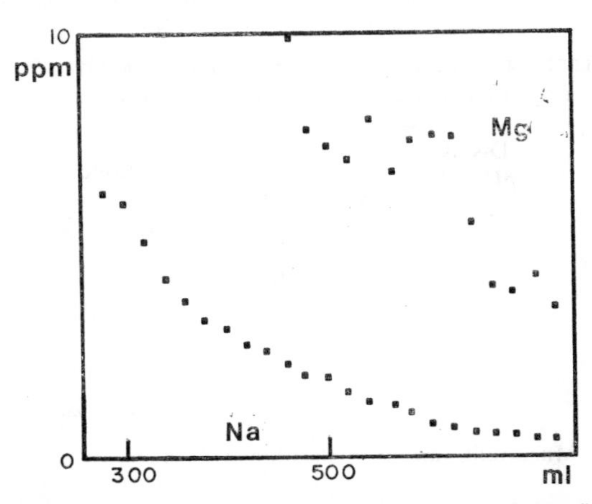

FIG. 5.2. Washing curves (metal ion concentration in ppm vs. water ml) recorded after passing a salt solution through a chitosan column (conditions listed in Table 5.1). (From R. A. A. Muzzarelli *et al.*, *J. Chromatogr.* **47**, 414 (1970).)

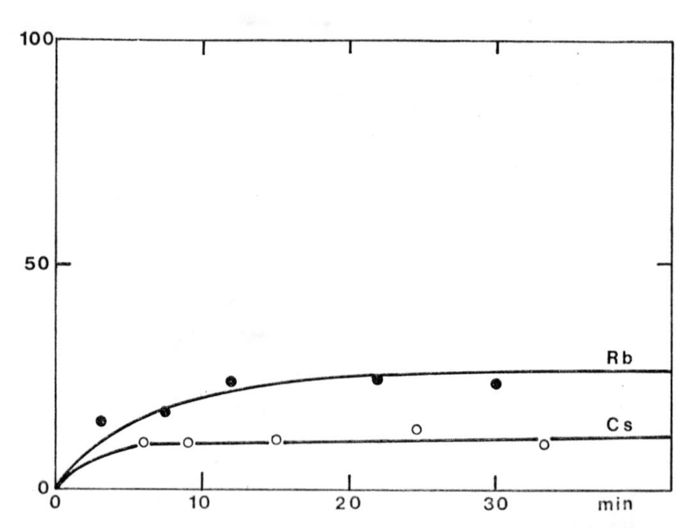

FIG. 5.3. Rates of collection of rubidium and caesium from 50 ml of 0·44 mM solution pH 6·0 by 200 mg chitosan. (From R. A. A. Muzzarelli *et al.*, *J. Radioanal. Chem.* **10**, 17 (1972).)

TABLE 5.2. COLLECTION % OF CAESIUM ON 200 mg OF 100–200 MESH POLYMER POWDER, AFTER 1 hr SHAKING WITH 50 ml SOLUTION, 0·44 mM, OR CARRIER-FREE

Concn.	Distilled pH = 7·0	Sea-water	Brine	Nitric acid pH = 2·0
Chitosan				
0·44 mM	12	0		0
c.f.	100	5	0	0
Chitin				
c.f.	57	3		0

(From R. A. A. Muzzarelli, various publications.)

FIRST-ROW TRANSITION METAL IONS COLLECTION ON CHITIN AND CHITOSAN

The collection percentages of metal ions on polymers are preliminary information that can be easily obtained and that can be quite useful for predicting the behaviour of the metal ions towards the polymers in column chromatography.

Table 5.3 presents the collection percentages of first-row transition and post-transition metal ions on chitin and chitosan powders 100–200 mesh. Chitosan exhibits the best collection ability for all polymers so far

TABLE 5.3. FIRST-ROW TRANSITION AND POST-TRANSITION METAL ION COLLECTION ON 200 mg CHITIN AND CHITOSAN. PER CENT OF THE AMOUNT OF METAL PRESENT IN 50 ml OF 0·44 mM AQUEOUS SOLUTION (ATOMIC ABSORPTION SPECTROMETRY)

	pH	hr	Cr(III)	Cr(VI)	Mn(II)	Fe(II)	Ni(II)	Cu(II)	Zn(II)	As(V)
Chitin	2·5	1	0	63	0	6	4	15	0	0
		12								
	EDTA	1	22	17	4	5	0	0	0	0
	5·5	1	40	12	12	52	8	38	22	0
		12							20	
	EDTA	1	20	8	20	0	9	0	0	0
Chitosan	2·5	1	17	81	21	28	28	2	12	100
		12	54	98	20	23	60	94	46	73
	EDTA	1	28	70	16	16	1	2	11	27
	5·5	1	38	12	9	79	72	86	71	25
		12		18	26	65	95	100	89	73
	EDTA	1	21	36	26	2	15	0	0	11

(R. A. A. Muzzarelli *et al.*, *Talanta* **16,** 1571 (1969); see also U.S. Patent 3,635,818 (1972).)

characterized as chelating polymers, as its collection ability is already evident after 1 hr contact time, especially for Ni, Cu, Zn and Cr(VI). The collection percentages are particularly low for Mn and Fe(II). When the solution is made 0·1 M EDTA, the latter prevents the collection of Ni, Cu and Zn, while for the other ions some collection takes place even under these conditions. This indicates that chitosan is a very powerful chelating agent, because of its high amino groups content.

Chitin, of course, does not exhibit a so high collection ability, as the nitrogen electrons of the acetylated amino group are not available as in chitosan.

Batch measurements extended over a certain period of time furnish the collection rate curves.

Several chromatographic results are in Table 5.4.

The curves of the collection rates depend on several factors, the most important of which is the polymer grain size, the temperature, stirring speed and mode, the presence of other ions suitable for collection, and oxidation level of the ion.

Grain size of the polymer powders

In Fig. 5.4 the collection rates of zinc on chitosan flakes (2 mm) and on 100–200 mesh powder are shown, the other conditions being similar. The powder collects zinc ions better, more rapidly and to higher extent than flakes, due mostly to the larger surface area immediately available for collection. This fact corresponds to well-known rules in chromato-

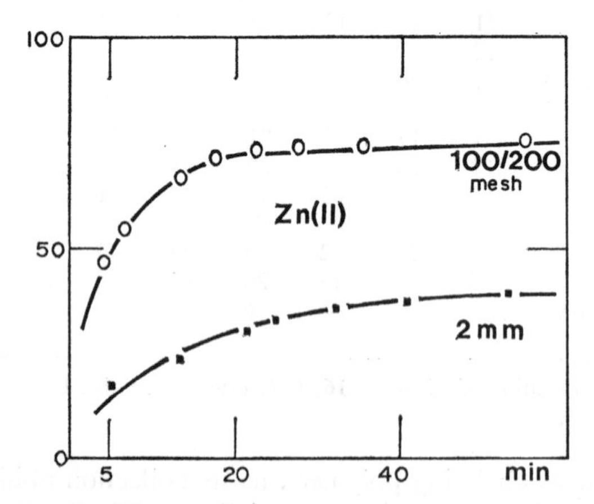

FIG. 5.4. Rates of collection of zinc from 50 ml of 0·44 mM solution by 200 mg chitosan powders having different grain sizes. (R. A. A. Muzzarelli, original results.)

graphy which state that the smaller the grains the smaller the polymer volume necessary for a chromatographic operation, and the equilibrium is reached more quickly.

As a chelating polymer has to be highly selective and give immediate collection, the collection rates are studied over a limited length of time, normally 1 hr, even though in a few cases the equilibrium is not reached.

Temperature

Temperature plays an important role in collection, and this appears from the positions of the curves in Fig. 5.5, which for lead is at 0 °C and 25 °C. There is no general rule for predicting the influence of the temperature on metal ion collection, and therefore one can expect variations in both directions.

Stirring mode and speed

In order to get reproducible results, when measuring distribution coefficients, stirring should be quite energetic, and applied in the same mode. Normally, a good magnetic stirrer is sufficient to provide

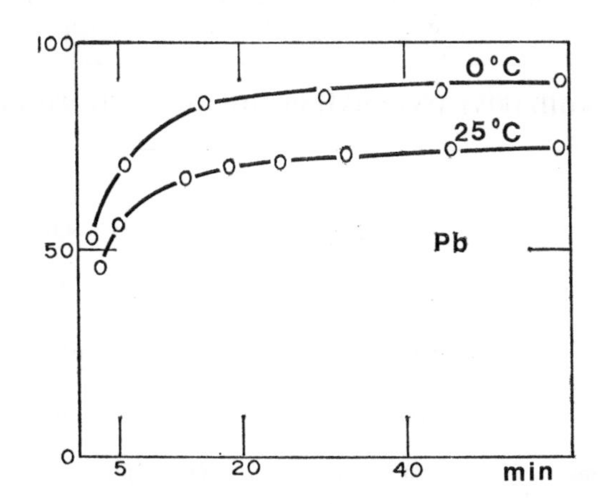

FIG. 5.5. Rates of collection of lead from 50 ml of 0·44 mM solution by 200 mg chitosan, at 0 °C and 25 °C. (R. A. A. Muzzarelli, original results.)

constant and effective stirring. However, ultrasonic stirring can alter the results considerably, as it is a completely different stirring mode. From Fig. 5.6, one can see that metal ion collection is effectively improved under ultrasonic stirring.

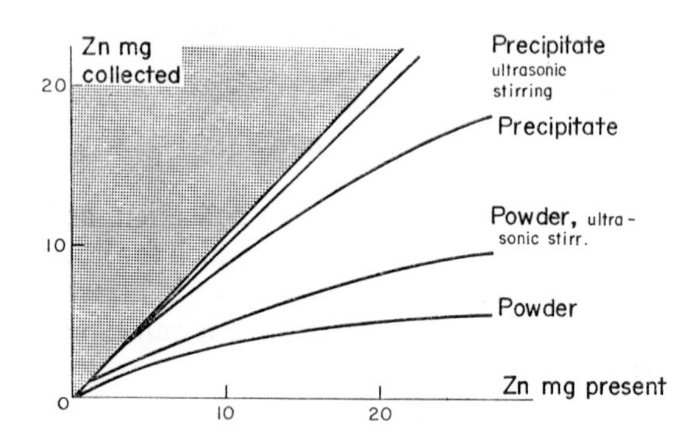

FIG. 5.6. Collection of zinc under various conditions, on 200 mg of chitosan, at pH = 6·6 and 20 °C. (mg Zn collected vs. mg Zn present in 50 ml.) (From R. A. A. Muzzarelli, *Anal. Chim. Acta* **54**, 133 (1971).)

TABLE 5.4. FIRST-ROW TRANSITION METAL IONS CHROMATOGRAPHY ON 7×1 cm COLUMNS OF 100–200 MESH CHITOSAN POWDER

	Sc(III)	Ti(IV)	V(V)	Cr(VI)	Mn(II)	Fe(III)	Co(II)	Ni(II)	Cu(II)	Zn(II)
Amount used μg in 100 ml				250			0·01	6000		0·02
Collection %	100		100	100		100	100	100	100	100
Eluent 0·1 N			NH_3	NaCl		EDTA	EDTA	EDTA	EDTA	EDTA
Elution %			100	100		100	89	100	100	79
Volume of eluent			20	20		60	60	100	60	60

Presence of other ions

The presence of Group I A and II A metal ions, and of ammonium and thallium(I) ions have no appreciable effect on the collection of transition metal ions.

Large amounts of ammonium sulphate allow transition metals, present in solution at the trace level, to be collected on chitosan columns. For instance, 100 ml of 0·44 M ammonium sulphate solution containing nanograms of radioactive zinc were passed through a 7×1 cm column of 100–200 mesh chitosan. In Fig. 5.7 (a), the radioactive profile

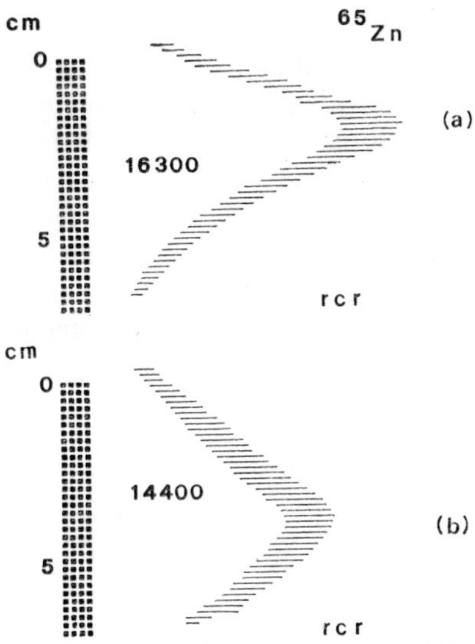

FIG. 5.7. Separation of traces of zinc from ammonium sulphate brines. Radioactive zinc distribution in a 100–200 mesh chitosan powder chromatographic column, (a) before washing, (b) after washing. (From R. A. A. Muzzarelli *et al.*, *J. Radioanal. Chem.* **10,** 27 (1972).)

of the column is reported; it shows that most of the radioactivity was fixed in the upper part of the column, and that zinc was completely collected. By washing with a further 200 ml of 0·44 M ammonium sulphate solution, the band is lowered a little and some zinc is eluted from the column. This is a very satisfactory test of the effectiveness of chitosan in separating trace transition metal ions from very high ionic strength solutions. Even sharper bands at the top of the column, were observed for copper. These data were obtained with a Desaga–Berthold chromatogram scanner equipped with an argon–methane gas-flow counter. The radiotracers in the column could be easily located with the columns mounted on the moving plate 1 mm below the detector window. For the figures reported here, the window was 1·2×0·3 cm, the time constant 1 sec, the pen excursion 30 c/s, speed 600 mm min^{-1}, and the integral readings were given by an automatic Metrawatt integrator. This technique is quite useful for understanding what is going on in the column, even during the chromatographic process because a suitable arrangement permits one to obtain readings while carrying out the chromatographic procedure.

As long as two or more transition metal ions are present in the solution under examination, along with a quantity of polymer, insufficient for the complete collection of both, the cation which forms the most stable complex with the polymer will be collected, leaving most of the other cation in solution. Some variations are introduced into these curves by the simultaneous presence of other cations. In any case, these variations are less than expected on the basis of a stoichiometric capacity.

Figure 5.8 shows the case of cadmium and nickel, the cations of which are strongly collected by chitosan. Their effect is reciprocal, and even slightly more evident for nickel.

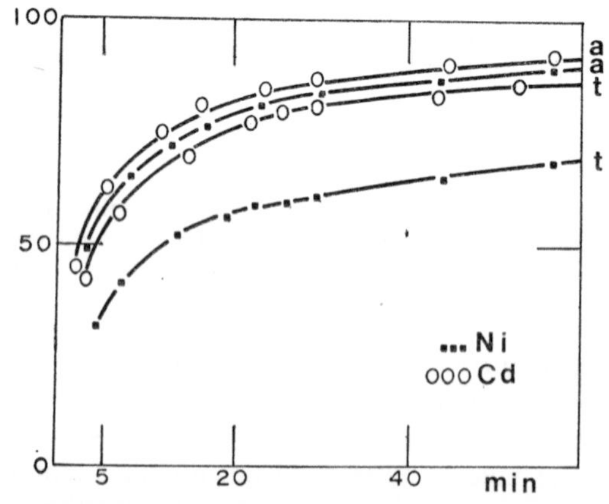

FIG. 5.8. Rates of collection of nickel and cadmium from 50 ml of 0·44 mM solution by 200 mg chitosan (t) together, (a) alone. (R. A. A. Muzzarelli, original results.)

When a cation is only collected to a smaller degree by chitosan, like iron(II), and it is in a solution of nickel which is well collected by chitosan, nickel appreciably depresses the collection of iron(II) as shown in Fig. 5.9. However, some iron can still be collected.

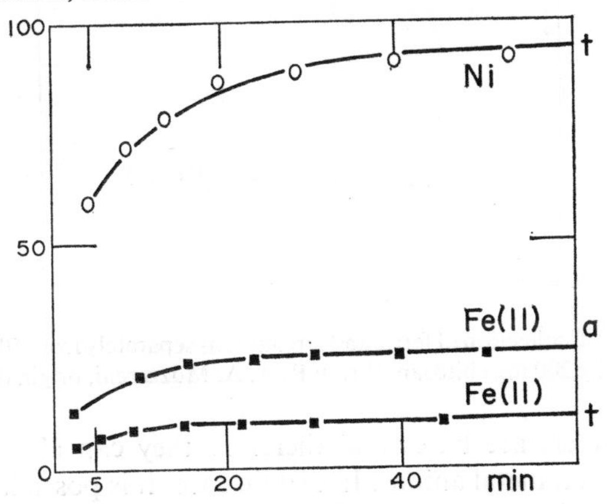

FIG. 5.9. Rates of collection of nickel and iron(II) from 50 ml of 0·44 mM solution by chitosan (t) together, (a) alone. (R. A. A. Muzzarelli, original results.)

Oxidation state of the cations

In Fig. 5.10, the different behaviour of ferrous and ferric ions is quite evident. This indication, as those above, is consistent with chromatographic results, as iron(II) can be easily eluted from chitosan columns, while iron(III) requires a preliminary reduction in order to be able to perform the elution.

It is interesting to compare the amount of iron(II) collected by chitosan to the amount of iron(II) collected by cellulose, as reported by Vial and coworkers.[5] From a solution containing 60 mg l^{-1}, 1 kg of cellulose collects 364 mg, while on the basis of the above reported data 1 kg of chitosan has a capacity of 3125 mg of iron(II), the experimental conditions being comparable, as the 200 mg polymer sample was used to treat 50 ml of 1 mM solution. Of course the capacity of chitosan for iron(III) is much higher.

The nitrogen electrons present in the amino and *N*-acetylamino groups can establish dative bonds with transition metal ions, especially in the case on chitosan, where free amino groups are particularly abundant.

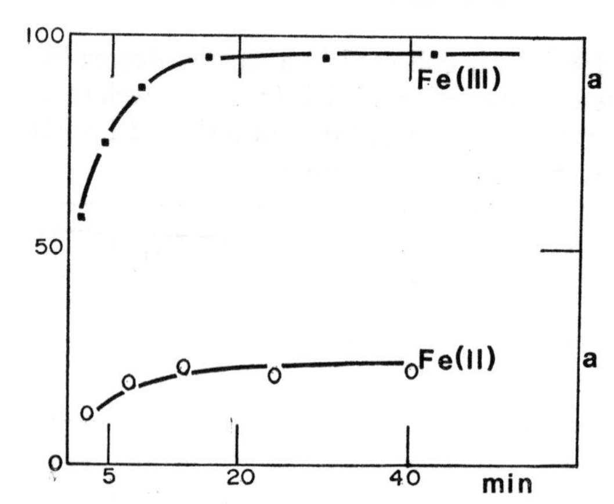

FIG. 5.10. Rates of collection of ferric and ferrous ions separately from 50 ml of 0·44 mM solutions by 200 mg chitosan. (From R. A. A. Muzzarelli, original results.)

These polymers are bases and therefore they can also act to form salts of transition metal anions. In certain cases it is possible that several types of interactions may occur simultaneously.

The interactions of the first-row transition metal ions with chitin and chitosan are accompanied by colour appearance in many instances, namely red for titanium, orange with metavanadate, green for chromium(III), and orange for chromium(IV), yellowish-brown with iron(II),

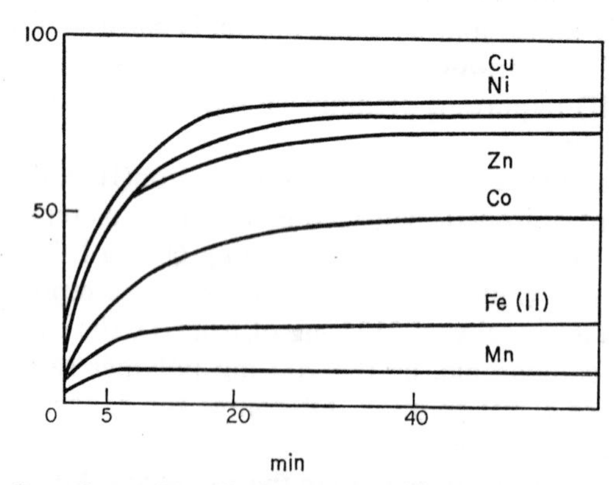

FIG. 5.11. Rates of uptake on Mn, Fe(II), Co(II), Zn, Ni and Cu(II) on 200 mg of chitosan, when taken separately in 100 ml of 0·4 mM solution (0·1 M KCl). (From R. A. A. Muzzarelli *et al.*, *Mikrochim. Acta* 892 (1970).)

yellowish-green with iron(III), pink with cobalt, green with nickel, and blue with copper. These colours are more intense with chitosan than with chitin. In certain cases, as with copper, the colour formation is quite sensible.

The sequence of the curves of the collection rates for the first-row transition metal ions in Fig. 5.11, is the same as the Irving & Williams Series,[6] and there is a relationship between this order and the second ionization potentials. Moreover, complexing agents are generally necessary to perform elutions from chitin and chitosan chromatographic columns. This experimental evidence is in favour of the chelating ability of chitin and chitosan, whose amino groups play a major role, as confirmed also by measurements carried out on dissolved chitosan.

Precipitation of transition metal ions by chelation

It is possible to use chitosan as a polymer for collection of trace metals by chelation accompanied by coprecipitation.[4]

Chitosan solutions were prepared as follows. Prepare a solution containing 10 mg ml^{-1} of chitosan powder in $1+9$ acetic or formic acid. Centrifuge at 20,000 rev min^{-1} for 15 min to eliminate undissolved particles, if any. Precipitate the polymer from the clear solution by adding dropwise and with vigorous stirring pure *ca.* 0·1 M NaOH solution. Lyophilize the precipitate after careful washing with water to neutral pH. The water should be free from any transition metal ion. Finally, dissolve for use a weighed amount of lyophilized chitosan to make a 10 mg ml^{-1} solution in $1+9$ formic or acetic acid.

When a solution of chitosan in dilute formic or acetic acid is poured into excess of water, chitosan precipitates as white flakes. Of course, the pH of the resulting solution is of great importance for the completeness of the precipitation. Chitosan solutions in formic acid were titrated with 0·1 M NaOH in order to establish visually the pH at which the precipitation begins. This pH value depends to a certain extent on the chitosan and metal ion concentrations, but it generally lies between pH 5·0 and 6·7.

The precipitation of chitosan in solutions containing metal ions was studied. To demonstrate that chelation takes place, various metal ions were selected. The hydroxides of these metals precipitate at a pH high enough to avoid any risk of precipitating a mixture of hydroxide and chitosan, instead of chitosan with fixed metal ions. The metals and the pH intervals for the precipitation of their hydroxides were as follows: zinc (6·8–13·5), cerium (7·3–14), lead (7·2–13·0), nickel (7·4–14), cobalt (7·5–14), silver (8·0–14), and cadmium (8·3–14).

When the titration curves are recorded, one can observe that chitosan precipitates in a rather narrow pH interval when alone in solution. The equivalence point is reached in correspondence with the theoretical value for formic acid as can be seen in Fig. 5.12 for 0·5 g of chitosan in 20 ml of 0·3 M formic acid.

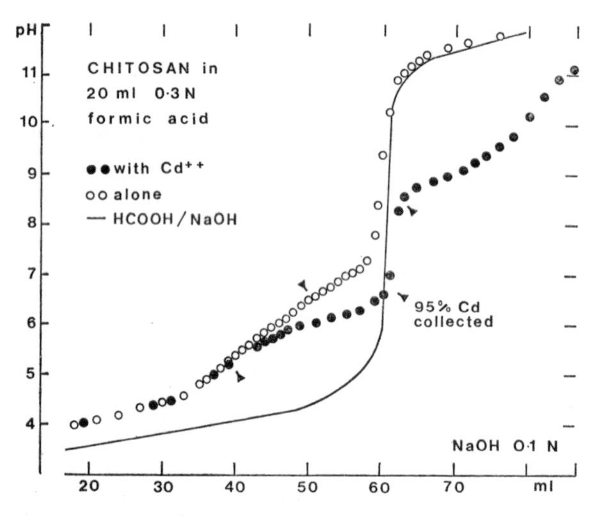

FIG. 5.12. Neutralization curves for (○) solution of 0·5 g of chitosan in 20 ml of 0·3 M formic acid; (●) the same with 0·25 g of $CdCl_2$ added; (——) formic acid alone. The arrows (from left to right) indicate the points at which chitosan + cadmium precipitate; the point at which chitosan alone precipitates; and the point at which 95% of cadmium is collected when ultrasonic stirring is used and the point at which cadmium hydroxide precipitation begins. (From R. A. A. Muzzarelli, *Anal. Chim. Acta* **54**, 133 (1971).)

The pH at which chitosan precipitates when cadmium ions are present is lower than for chitosan alone (see arrows in Fig. 5.12). The equilibrium in formic or acetic acid solutions

$$—NH_2 + HCOOH \xrightleftharpoons{pH\ 3-4} —NH_3^+ + HCOO^-$$

is shifted to the right and the nitrogen electrons become unavailable for complexing cadmium(II). Polarographic waves for 2 ml of 0·01 M cadmium solution diluted in one test with 4 ml of 1 M acetic acid, and in a second with 4 ml of 1 M acetic acid containing 40 mg of chitosan, were identical, with the half-wave potential at −0·65 V. The same was true for lead in perchloric acid solutions, the half-wave potential occurring at −0·50 V in both cases. This finding is consistent with light-scattering data obtained on chitosan solutions where chitosan was found in the $—NH_3^+$ form.

Apparently, at higher pH the equilibrium

$$—NH_3^+ + Cd^{2+} \rightleftharpoons —NH_2 \rightarrow Cd^{2+} + H^+$$

is shifted to the right so that significant interaction of cadmium ions with chitosan occurs, and up to 95% of the cadmium present is collected on the precipitating chitosan. Light-scattering measurements confirmed that the polymer is polydispersed with an average molecular weight of 120,000. The rather broad shoulder in Fig. 5.12 indicates that there is some fractionation during precipitation.

Precipitation of chitosan generally improves the collection yields compared with addition of chitosan powder (Table 5.5), largely because of the increased uptake during the first minutes. For stirring times longer than 1 hr, the metal ion capacity seems to be about the same for chitosan powder, lyophilized chitosan and precipitated chitosan under the same conditions. When chitosan powder is used, the metal ions have to penetrate into the polymer particle and the collection rate is lower than for precipitating chitosan, the nascent flakes of which offer a very large surface area for reaction.

The effects of ultrasonic stirring on collection yields are also shown in Table 5.5. In Fig. 5.6 the amount of zinc(II) collected by chitosan is plotted as a function of zinc(II) concentration; this shows the improved uptake on precipitating chitosan under mechanical and ultrasonic stirring. Similar curves were obtained for cobalt.

When precipitated in dilute ($2{\cdot}4 \times 10^{-7}$–$8{\cdot}3 \times 10^{-4}$ M) metal ion solutions, chitosan collects metal ions with good yields. In Table 5.6 data are presented for collection 10 and 20 min after precipitation under standard stirring conditions at various pH values. These results show that precipitated chitosan forms chelates with metal ions in slightly acidic solutions, and at low enough pH values that hydroxides are not

TABLE 5.5. PER CENT COLLECTION OF TRANSITION AND POST-TRANSITION ELEMENTS FROM 0·44 mM SOLUTIONS BY PRECIPITATION OF 200 mg OF CHITOSAN (FINAL pH BETWEEN 6·5 AND 7·1, ADJUSTED WITH 0·1 N NaOH OR NH_4OH. RADIOCHEMICAL AND POLAROGRAPHIC MEASUREMENTS)

Element	Mechanical stirring 1 hr		Ultrasonic stirring 15 min
Iridium	30	(15)*	30
Cobalt	60	(40)	83
Silver	100	(100)	100
Zinc	70	(61)	100
Mercury	100	(90)	100
Cadmium	81	(77)	95

* In parenthesis are the corresponding values for chitosan powder.

(From R. A. A. Muzzarelli, *Anal. Chem. Acta* **54,** 133 (1971).)

TABLE 5.6. COLLECTION OF METAL IONS BY 100 mg OF CHITOSAN PRECIPITATED IN DILUTE SOLUTIONS AS A FUNCTION OF CONCENTRATION (LESS THAN 0·2 mM) AND OF pH, AFTER MECHANICAL STIRRING FOR 10 OR 20 min (ATOMIC ABSORPTION MEASUREMENTS)

Metal ion concn. (μg ml^{-1})				% Metal ion collected after	
Original soln.	After 10 min	After 20 min	pH	10 min	20 min
Cadmium (chloride)					
5	0·80	0·80	7·0	84	84
	0·70	0·70	7·5	86	86
	0·50	0·50	8·3	90	90
10	2·00	1·80	7·0	80	82
	1·42	1·40	7·5	86	86
	1·00	1·00	8·3	99	99
50	5·80	5·40	7·0	88	89
	4·52	4·50	7·5	81	81
	3·00	3·00	8·3	94	94

TABLE 5.6 *(cont.)*

Metal ion concn. (μg ml^{-1})				%Metal ion collected after	
Original soln.	After 10 min	After 20 min	pH	10 min	20 min
Cobalt (chloride)					
5	4·0	3·8	6·6	20	24
	3·4	3·2	7·0	32	36
	3·0	2·9	7·5	40	42
10	6·0	6·0	6·6	40	50
	5·1	5·1	7·0	49	49
	5·0	4·8	7·5	50	52
50	22·0	22·0	6·6	56	56
	20·5	20·5	7·0	59	59
	19·0	19·0	7·5	62	62
Nickel (chloride)					
5	3·00	3·00	6·6	40	40
	0·93	0·92	7·0	81	84
	0·94	0·92	7·4	81	82
10	4·80	4·80	6·6	52	52
	2·10	2·00	7·0	79	80
	1·70	1·70	7·4	83	83
50	22·00	22·00	6·6	56	56
	11·00	11·00	7·0	78	78
	5·00	4·80	7·4	90	90
Lead (nitrate)					
5	1·90	1·70	6·6	62	66
	1·50	1·20	6·9	70	76
	0·50	0·37	7·2	90	93
10	4·40	3·80	6·6	56	62
	3·00	2·90	6·9	70	71
	2·50	2·30	7·2	75	77
50	22·0	20·1	6·6	56	60
	20·0	19·0	6·9	60	62
	18·0	17·0	7·2	64	66
Zinc (chloride)					
5	2·90	2·80	6·6	42	44
	1·80	1·25	6·8	64	75
10	5·6	5·2	6·6	44	48
	3·0	2·0	6·8	70	80
50	21·0	19·0	6·6	58	62
	10·0	9·80	6·8	80	80

(From R. A. A. Muzzarelli, *Anal. Chim. Acta* **54**, 133 (1971).)

precipitated. Should the pH be raised, chitosan would also act as a carrier for the metal hydroxides, which would be difficult to separate at these low concentrations without the help of a coprecipitation agent. These measurements also indicate that precipitated chitosan has rather low rates of uptake for certain metal ions.

When chitosan is precipitated in a solution where two or more transition or post-transition metal ions are present, competitive reaction occurs and the amount of collected metal ions is slightly reduced. In Table 5.7 values are reported for nickel and cobalt on chitosan powder and precipitate.

TABLE 5.7. COMPETITIVE REACTION OF TWO IONS IN COLLECTION ON CHITOSAN (100 mg) AT pH 7·0 WITH MECHANICAL STIRRING (ATOMIC ABSORPTION MEASUREMENTS; 50 ml OF METAL ION SOLUTION)

Metal ion concn. (μg ml^{-1})			% Metal ion collected after			
			10 min		20 min	
Original soln.	After 10 min	After 20 min	Precip.	Powder	Precip.	Powder
Cobalt						
5	3·4 (3·4)*	3·2 (3·2)	32 (32)		36 (36)	
10	5·4 (5·1)	5·4 (5·1)	46 (49)		46 (49)	
50	26 (21)	24 (21)	48 (59)	13 (40)	52 (59)	13 (47)
Nickel						
5	1·3 (0·9)	1·3 (0·8)	74 (81)		74 (84)	
10	2·7 (2·1)	2·4 (2·0)	73 (79)		76 (80)	
50	12·5 (11)	11·5 (11·0)	75 (78)	31 (65)	78 (78)	31 (78)

* The values for the element taken separately are given in parentheses.

(From R. A. A. Muzzarelli, *Anal. Chim. Acta* **54,** 133 (1971).)

Co-precipitation of transition metal compounds

A process for colouring glass fabrics and other fibres has been patented recently.[9, 10] The process of this invention involves mixing a cationic flocculating agent with a dispersion of a pigment, to produce a dispersion containing flocculated pigment particles, and treating the fabric with this dispersion. While anionic pigment dispersions are preferred, any pigment which can be charged anionically may be employed. Pigments that are useful in practice are the inorganic pigments, chiefly the oxides, sulphites and sulphates of cobalt, chromium, aluminium, iron, zinc, cadmium, manganese, and selenium, and organic pigments such as the phthalocyanine dyes; anthraquinone, thioindigoid, and indanthren vat dyes; azo coupling dyes; Hansa yellow and carbon black.

Chitosan is the preferred cationic flocculating agent employed in the dyeing bath. While it is suggested to use acetic acid for its preparation, other acids can be used, namely formic, pyruvic and lactic acids. The detailed preparation of the latter salts of chitosan is described in other patents. The cationic flocculating agent is included in the bath in concentrations of 0·005 to 5% by weight, preferably in the region 0·05 to 1%. The pH of the bath is between 9·0 and 10·5 and ammonia is used to control it. Baths with a pH above 5·5, however, have been found satisfactory, and in fact this is the pH of chitosan flocculation. The bath also contains a resinous film-forming binder for bonding the pigments to the fibres of the fabric, such as butadiene-acrylonitrile or butadiene-styrene resins, in the range 0·5 to 2% by weight, based on the weight of the bath. Other substances can be added as well, like colloidal silica for glass fibres.

The flocculating polymer chitosan is essential in this process, because the flocculated pigment particles dye the loosely constructed areas of the fabric in preference to, or to the exclusion of, the tightly constructed areas.

For example a bath was composed of 1% yellow iron oxide in 10% water, 1% chitosan solution, 0·5% melamine–formaldehyde resin and about 87·5% methanol. The dispersion of the dye in water was placed in a container, methanol and chitosan were added under mixing until agglomeration resulted and then the resin was added to the resulting bath. By padding an all-glass fibre fabric having bulked yarn, through this dye bath and curing the fabric, a two-tone fabric resulted with bulked yarns coloured a bright yellow and the background zones an off-white shade.

For this industrial process, chitosan solutions have been found superior to polyalkyl–polyamines, polyalkyl–polyamine–fatty acids reaction products, amino aldehyde condensate dye-fixatives, dicyandiamine formaldehyde dye fixatives and many other flocculating agents, because it was able to effect the flocculation of a substantial portion of the pigment particles to maintain the solid constituents uniformly dispersed throughout the bath, and therefore to produce contrasting tonal effects.

Chitosan columns for chemical oceanography

The technique based on the precipitation of chitosan for the selective collection of transition and post-transition metal ions can be a useful analytical tool. The rather low pH for precipitation is advantageous because the final solution is almost neutral, and the precipitation of magnesium hydroxide is avoided when dealing with natural waters.

In any case, the importance of the research carried out on the precipitation of this polymer is, in the fact, that it helped to establish a procedure for the preparation of a powder having a very large surface area. In fact, the precipitate can be lyophilized, and the powder so obtained is quite suitable for a very rapid chromatography of dilute solutions, especially natural waters. With the precipitated chitosan powder, narrower bands and more efficient separations can be achieved while keeping high flow-rates.

An important application of chitosan in this physical form in column chromatography, is the collection of naturally occurring copper, zinc, nickel, cadmium and lead from sea-water.[7, 8] In fact, after extended trials with powders and precipitates, a filter unit was designed having the dimensions of 3×1 cm, made with precipitated and lyophilized chitosan. As indicated previously, this powder has a very large surface and is immediately and completely accessible to the dissolved chemical species. This cartridge is good for passing 10 l sea-water at a flow rate around 10 ml min^{-1}; it carries attachments to permit a counterflow elution of the collected metals. Elution can be generally carried out with a peristaltic pump and a few ml of a solution of complexing agent, and a volume reduction of 2000 is easily accomplished.

To define the behaviour of chitosan in sea-water from the standpoint of its ability to collect naturally occurring quantities of transition and post-transition metal ions, anodic stripping voltammetry offers excellent possibilities for several elements. A composite mercury–graphite electrode (CMGE), whose description has been recently published, allows one to determine very low concentrations of several elements in sea-water, among which are zinc, cadmium, lead, and copper. The determination of cadmium and lead by this technique is particularly valuable because these elements cannot be measured by neutron activation analysis.

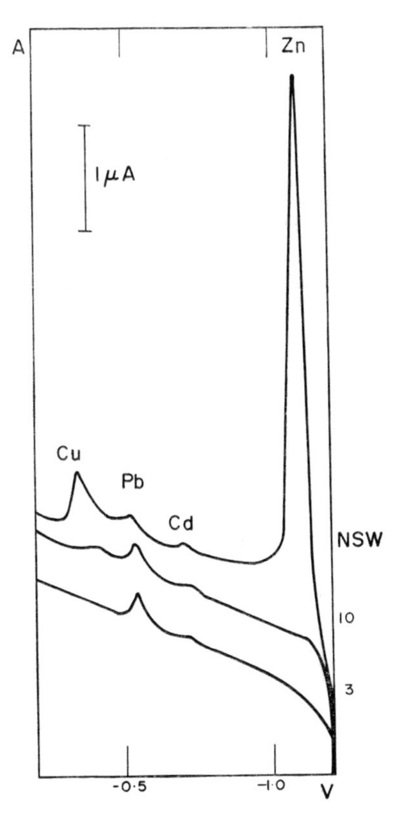

FIG. 5.13. Anodic stripping voltammetric curves from untreated (NSW curve) sea-water, and for the same water after passage through a 5×1 cm chitosan column at 10 ml min^{-1} and at 3 ml min^{-1} (100–200 mesh powder). (From R. A. A. Muzzarelli *et al.*, *Talanta* **18,** 853 (1971).)

By this technique, the four elements were determined on the water before and after passage through the chitosan columns at various flow-rates. The so obtained anodic stripping voltammetric curves are in Fig. 5.13. This figure is for a chitosan powder column, which is not so fast as a chitosan lyophilized powder column: in fact with the former a 3 ml min^{-1} flow-rate should be preferred, while with the second a 10 ml min^{-1} flow-rate can be currently adopted.

It can be seen that copper and zinc are very efficiently collected by chitosan. The elution of the mentioned elements can be performed according to following specifications: the best eluent sequence with solutions which are also suitable as electrolytes in voltammetry is: elute 100% Pb+ 100% Zn with 50 ml of 2 M ammonium acetate; elute 100% Cu with 10 ml of 0·01 EDTA and finally elute 100% Cd with 3 ml 0·1 M KCN. Of course, these reagents should be very pure: their impurity level limits the applicability of this procedure to small sea-water samples.

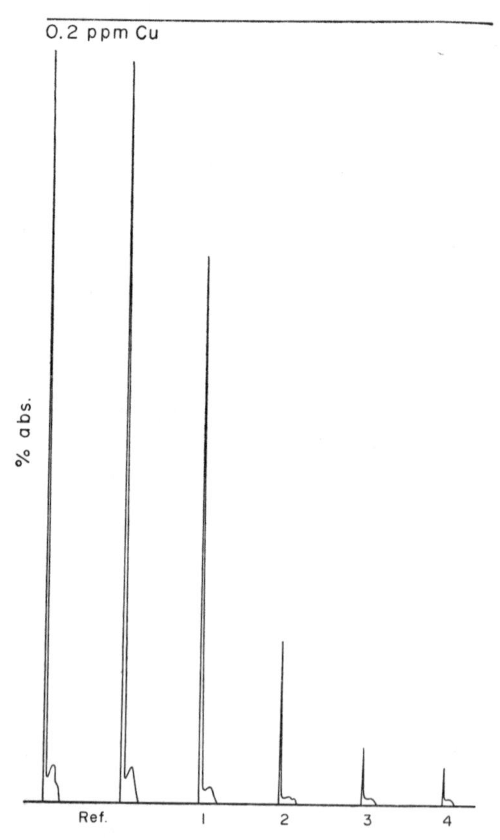

FIG. 5.14. Hot graphite atomic absorption determination of copper on four 5-ml eluate fractions, containing 1,10-phenanthroline.

(R. A. A. Muzzarelli, original results.)

Another procedure has been devised for the elution of copper with 1% solution of 1,10-phenanthroline: this reagent is very pure by respect to copper and permits the quantitative determination of copper by hot graphite atomic absorption spectrometry (HGAAS): copper can be easily determined on 100-ml aliquots of sea-water by using 1·5×0·3 cm chitosan columns. The readings relevant to the four 5-ml eluate fractions are presented in Fig. 5.14. This method can be favourably compared to a method based on the chelating resin Dowex A-100, as summarized in Table 5.8.

TABLE 5.8. COMPARISON OF METHODS FOR THE DETERMINATION OF COPPER IN SEA-WATER

	Riley and Taylor*	Muzzarelli and Rocchetti†
Polymer	Dowex A-100, 50–100 mesh	Chitosan, 100–200 mesh
Column size, cm	6·0×1·2	1·5×0·3
Sea-water volume, ml	1000	100
Washing, ml	250	15
Eluents	30 ml HNO_3 2 N	20 ml 1% 1,10-phenanthroline
Elution temperature	room	≈50 °C
Supplementary manipulations	evaporate to dryness, add 1 ml 0·1 N HNO_3 add 9 ml acetone	none
Analytical technique	Atomic absorption spectrometry	Hot graphite atomic absorption spectrometry

* J. P. Riley and D. Taylor, *Anal. Chim. Acta* **40**, 479 (1968).
† R. A. A. Muzzarelli and R. Rocchetti (original results).

Vanadium and molybdenum can be collected on chitosan from acidified sea-water (pH = 2·5). Vanadium can be eluted with 0·1 N ammonia and determined by HGAAS. In the light of the data presented in Table 1.2, DE and PAB celluloses are also suitable for the collection of molybdenum from acidified sea-water. Collection is in fact 100% on 3×1 cm columns of the three polymers, and the best elution yield was reached with PAB cellulose, as shown in Fig. 5.15, by eluting with 1 N ammonium carbonate.

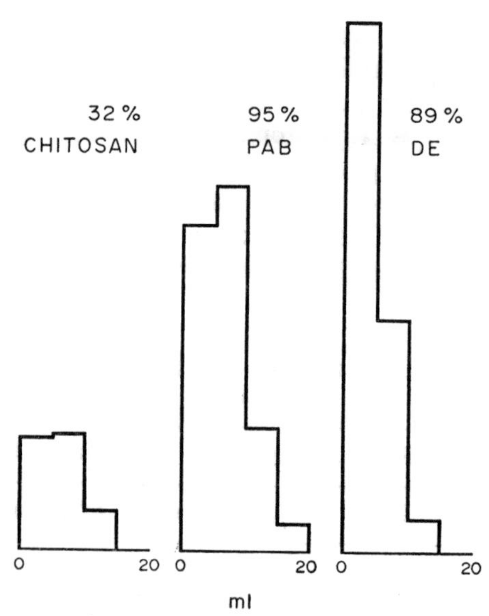

FIG. 5.15. Elution curves of molybdenum from 3×1 cm columns, with 1 N ammonium carbonate

(From R. A. A. Muzzarelli and R. Rocchetti, *Anal. Chim. Acta*, to appear in 1973).)

The results obtained with the addition method, the reading proportionality for increasing amounts of sea-water, and their reproducibility for aliquots of the same sea-water sample confirmed the validity of this new method, which compares favourably with other methods, as shown in Table 5.9, and which has been developed by combining the high selectivity of the natural chelating polymer with the extreme sensitivity of the HGAAS.

In so far as zinc, cadmium, and copper are discharged to the sea in ionic form, which can be easily collected on chitosan, the collection methods described here seem to be susceptible in the application of the marine pollution survey field. As far as clear sea-water is concerned, it seems that chitosan can be used for preparation of low metal ion content sea-water, for biological studies.

The rates and the yields of uptake of chitosan are very high, so once standardized, the method based on chitosan would be useful for rapid preconcentration treatment of sea-water on very small filters, precoated

TABLE 5.9. A COMPARISON OF METHODS FOR THE DETERMINATION OF MOLYBDENUM IN SEA-WATER

	Muzzarelli and Rocchetti*		Riley and Taylor !	Kawabuchi and Kuroda §	
Polymer amount, g	Chitosan 0·5	PAB Cell. 0·5	Dowex A-100 –	Dowex 1-X8 10·0	Dowex 1-X8 10·0
Column conditioning dimensions, cm	none 1×3	none 1×3	nitrate 1×6	chloride 2·5×3·5	thiocyanate 2·5×3·5
Sea-water volume, ml	50	50	3000	1000	1000
pH adjustment	2·5	2·5	5·0	0·1 M HCl+6% H_2O_2	0·1 M HCl+0·1 M SCN
Washing, ml	15 water	15 water	200 water	50, 0·5 M NaCl	100, 1 M H_2SO_4
Eluate, ml	none	10 NH_4CO_3	24 NH_4OH 2 N	60, 0·5 M NaOH +0·5 M NaCl	60, 0·5 M NaOH +0·5 M NaCl
Further treatments (extractions, etc.)	none	none	required	required	required

* R. A. A. Muzzarelli and R. Rocchetti, *Anal. Chim. Acta*, in press.
! J. P. Riley and D. Taylor, *Anal. Chim. Acta* **41**, 175 (1968).
§ K. Kawabuchi and R. Kuroda, *Anal. Chim. Acta* **46**, 23 (1969).

filters, cartridges or columns, to avoid storage, acidification and transportation of water samples.

REACTIONS WITH ZIRCONIUM, HAFNIUM, AND NIOBIUM

Chitosan can be precipitated from its solutions, by solutions of zirconium tetrachloride at a strongly acidic pH. The precipitate is white and brilliant and forms immediately. This indicates a quite important and exceptional interaction of zirconium with chitin and chitosan.[11]

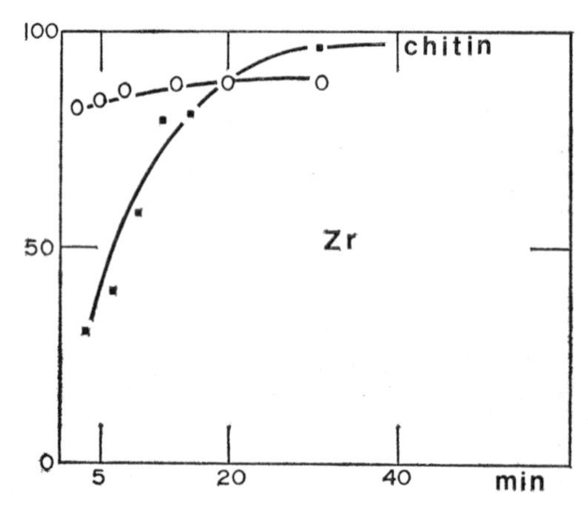

FIG. 5.16. Rates of collection of zirconium ions from 50 ml of 1 mM solution on 200 mg chitosan and chitin, 100–200 mesh. (From R. A. A. Muzzarelli *et al.*, *J. Radioanal. Chem.* **10**, 17 (1972).)

In fact, zirconium is collected on chitin and chitosan powders not only from chloride solutions, but also from thenoyltrifluoroacetone solutions (Figs. 5.16, 5.17): the latter is a very strong complexing agent for zirconium, but, nevertheless zirconium is retained on the column in a very narrow band at the top. When cadmium and zirconium are together in a chloride solution, the latter depresses remarkably the cadmium collection, as can be seen in Fig. 5.18.

Zirconium chelates were proposed as crosslinking agents for cellulose acetate,[12] and may be zirconium partially acts this way also with chitin and chitosan. Oxidized cellulose was used to prepare chromatographic columns through which 100 ml of zirconium sulphate or chloride solutions were passed at the flow-rate of 3–4 ml min^{-1}. Collection was

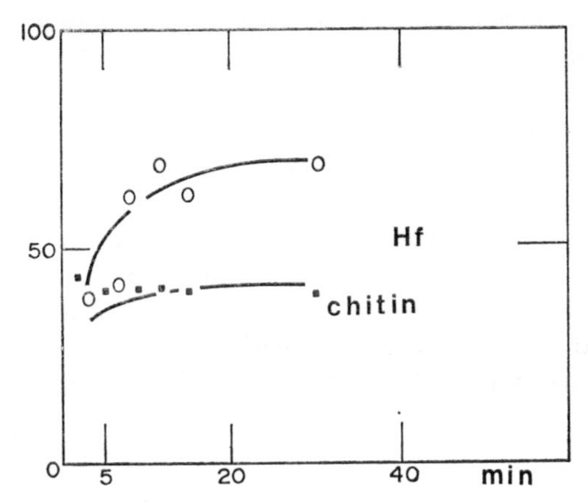

FIG. 5.17. Rates of collection of hafnium ions from 50 ml of 1 mM solution on 200mg chitosan and chitin, 100–200 mesh. (From R. A. A. Muzzarelli *et al.*, *J. Radioanal. Chem.* **10**, 17 (1972).)

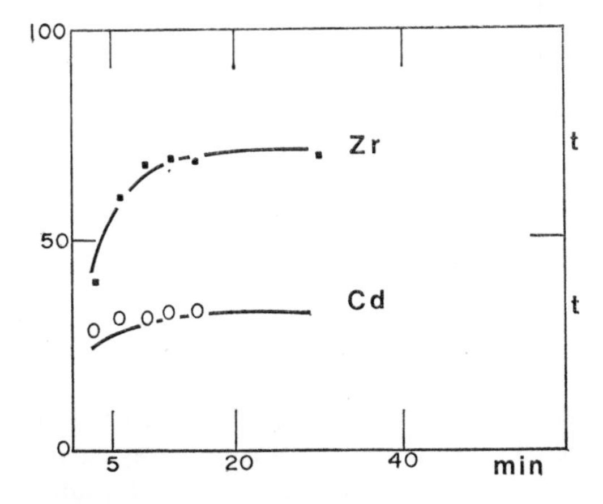

FIG. 5.18. Rates of collection of zirconium and cadmium ions from 50 ml of 1 mM solution on 200 mg chitosan, 100–200 mesh, when present together. (R. A. A. Muzzarelli, original results.)

between 93 and 100%, the highest value being for sulphate solution at pH 1·7–1·8. Desorption was carried out with 20 ml of 3·5 N HCl or 5–6 N sulphuric acid at 60–70 °C. The reported capacity of oxidized cellulose was 0·29 meq Zr g^{-1}.[13] The existing information on cellulose and zirconium seems to indicate that a main interaction exists with the anhydroglucose polymer chain but in the case of chitin and chitosan an interaction occurs with the nitrogen atoms also. Several authors studied this kind of interaction.[14]

The solution acidity can play a certain role in the collection of zirconium and other ions on chitin and an optimum pH, around 2, should be selected for the best retention; as at this pH no caesium is collected on the column, one can therefore deduce that the separation of caesium from zirconium and other fission products would be feasible, particularly when high acidity solutions are dealt with.

It is worth mentioning that titanium is poorly collected, as shown in Fig. 5.19.

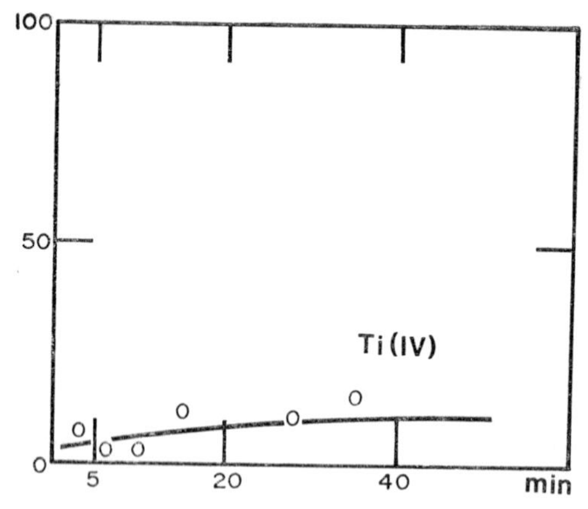

FIG. 5.19. Rate of collection of Ti(IV) ions from 50 ml of 0·44 mM solution by 200 mg of chitosan. (R. A. A. Muzzarelli, original results.)

^{137}Cs is ideally suited for burn-up determinations because of its nuclear characteristics. However, as its gamma-ray energy is lower than the ^{144}Ce, ^{95}Zr and ^{95}Nb gamma-energies, the photopeak of ^{137}Cs cannot be measured accurately because of the underlying contributions due to the cited radioisotopes. Therefore, in the light of the previously described caesium behaviour, chitin and chitosan have been found suitable for a clean separation procedure.[11] Chitin column data are presented in Table 5.10 and it can be observed that zirconium and niobium can be separated from the other fission products.

From the data in Table 5.11, one can foresee that an even sharper separation, i.e. Cs from Ru, Ce, Zr, and Nb, can be performed when

TABLE 5.10. PER CENT OF THE METAL ION QUANTITY FOUND IN SOLUTION, AFTER CHROMATOGRAPHY OF 100 ml ON 1×7 cm CHITIN COLUMN

Element	pH 1·0	pH 2·0	pH 2·5	pH 3·0
Ruthenium	100	100	100	76
Cerium	100	100	100	100
Caesium	100	100	100	100
Zirconium+Niobium	45	15	0	0

(From R. A. A. Muzzarelli *et al.*, *J. Radioanal. Chem.* **10**, 17 (1972).)

TABLE 5.11. PER CENT OF THE METAL ION QUANTITY FOUND IN SOLUTION AFTER OXIDATION WITH PERSULPHATE, AND CHROMATOGRAPHY OF 100 ml ON 1×7 cm COLUMN

Element	Chitin	Chitosan
Ruthenium	30	5
Cerium	20	0
Caesium	100	100
Zirconium+Niobium	0	0

(From R. A. A. Muzzarelli *et al.*, *J. Radioanal. Chem.* **10**, 17 (1972).)

passing the fuel solution through a chitosan column, after oxidative treatment. In practice, a perfect separation as presented in Fig. 5.20 can be achieved on chitosan column. The results reported above have been extended to low level activity waters from nuclear plants. The results for

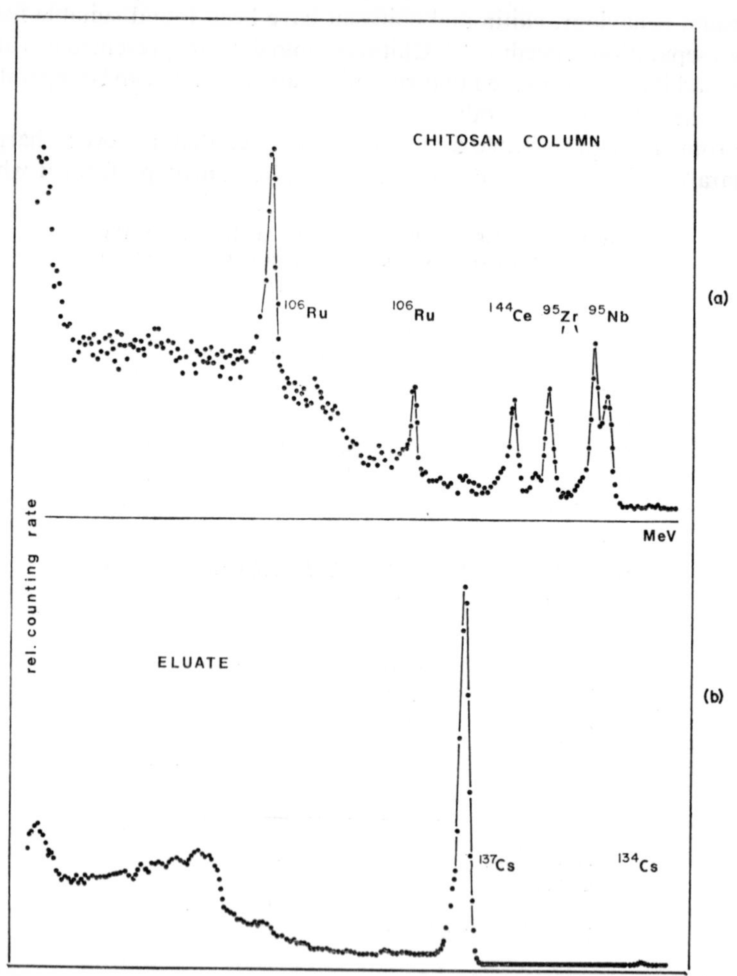

FIG. 5.20. γ-ray spectra of a chitosan column fraction (a) and of the eluate fraction (b) after passage of a fuel solution submitted to oxidation. (From R. A. A. Muzzarelli *et al.*, *J. Radioanal. Chem.* **10**, 17 (1972).)

these waters, after chromatography on chitin or chitosan are reported in Table 5.12. It may be observed, that the trend is the same as in the other cases, and there is agreement with the data reported on ruthenium.

TABLE 5.12. CHROMATOGRAPHY ON CHITIN AND CHITOSAN OF WASTE WATERS, pH = 3, IN mCi/m^3

Radioisotope	Before chromato-graphy	Chitin chromato-graphy	Chitosan chromatography		
			Direct	$KMnO_4$	$KMnO_4$ $+Na_2S_2O_8$
$^{95}Zr+Nb$	2·4	1·1	0·33	~0·1	~0·1
^{103}Ru	3	2	1·8	~0·8	~0·2
^{106}Ru	23	18	13	6	1·4
^{137}Cs	3	~2·6	3	3	3
^{134}Cs	0·5	0·5	0·5	0·5	0·5
^{144}Ce	~3·5	~3·5	<1	<1	<1
^{141}Ce	0·5	0·2	0·2	0	0
^{125}Sb	1	1	1	~0·5	~0·3

(From R. A. A. Muzzarelli *et al.*, *J. Radioanal. Chem.* **10,** 17 (1972).)

CHITOSAN SALTS AND DERIVATIVES IN CHROMATOGRAPHY

A few metals which form oxyanions, substantially deviate from the behaviour of the metal ions described above. Several oxyanions were tested, particular attention being given to sodium metavanadate, potassium dichromate, sodium molybdate, ammonium paramolybdate, and sodium tungstate. These salts were dissolved in dilute formic or acetic acid, and then mixed with chitosan in formic or acetic acid. Common practice was to add, dropwise, the oxyanion solution to the chitosan solution with very vigorous stirring. Immediate precipitation occurred to yield the inorganic derivatives of chitosan.[4]

In order to check the effectiveness of the collection of molybdate or tungstate, several solutions at various concentrations were treated with chitosan solution (200 mg of polymer, final volume 500 ml, stirring time 15 min, pH 3·0–4·0). The data obtained (Table 5.13) show that almost complete collection of molybdate is achieved in all cases, whereas collection of tungstate is less satisfactory. For metavanadate and chromate, there is a threshold around the metal concentrations 250 and 500 μg

TABLE 5.13. COLLECTION OF MOLYBDATE AND TUNGSTATE BY ADDITION OF 200 mg OF CHITOSAN IN FORMIC ACID TO SODIUM MOLYBDATE OR SODIUM TUNGSTATE SOLUTIONS AT pH 3–4

(Final volume 50 ml, 15 min stirring after precipitation; atomic absorption measurements)

Metal ion concn. (μg ml^{-1}) Before	After precipitation	% Collection	Metal ion concn. (μg ml^{-1}) Before	After precipitation	% Collection
Molybdenum			Tungsten		
2500	36·0	98·5	1000	270·0	73
1000	16·0	98·4	500	170·0	66
500	18·0	96·4	125	33·8	73
125	6·5	94·8	50	24·5	51
50	3·0	94·0			

(From R. A. A. Muzzarelli, *Anal. Chim. Acta* **54,** 133 1971).)

ml^{-1}, respectively, for precipitation to occur at pH 3–4. If the pH were raised, collection should be possible below these thresholds. The data in Table 5.13 and the infrared spectra (Fig. 5.21) indicate that molybdate and tungstate ions are bound to the polymer chain; they react more vigorously than the other oxyanions mentioned, possibly because of salt formation and polymerization. These precipitates were found to be amorphous by X-ray diffraction. They have a fibrous aspect and show a

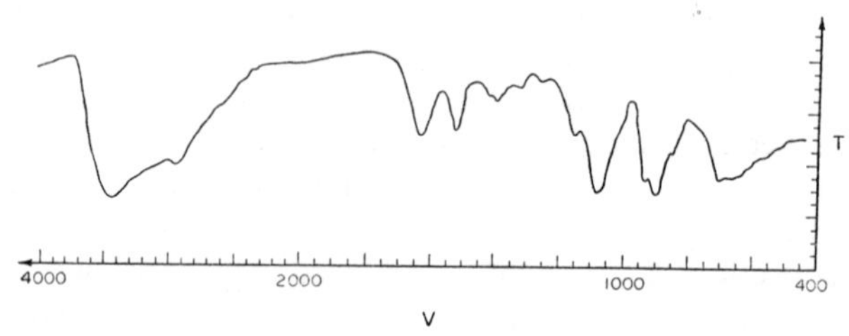

FIG. 5.21. Infrared spectrum of chitosan polymolybdate.

reversible photochromic effect when exposed to sun light: the metavanadate which is yellow, becomes green, the molybdate and tungstate which are white, become light blue, whilst dichromate, which is orange, undergoes oxidative degradation and becomes brown. A red derivative of chitosan obtained by reaction with exachloroplatinic acid (but not

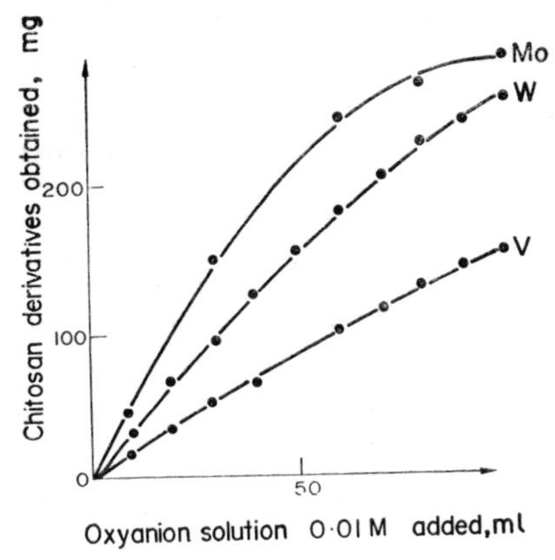

FIG. 5.22. Amounts of chitosan derivatives obtained when adding measured amounts of oxyanion to a 100 mg chitosan solution of acetic acid. The precipitates were lyophilized before weighing. (From R. A. A. Muzzarelli *et al.*, *J. Radioanal. Chem.* **10,** 27 (1972).)

with potassium hexachloroplatinate) has also been observed. These derivatives can be decomposed with concentrated carbonate solution.

The isolation of the molybdate and tungstate derivatives is not only important from the standpoint of a deeper insight in the retention mechanism of metal ions on natural polymers, but also for further developments in the application of these derivatives as chromatographic supports. In fact, derivatives of chitosan with metavanadate, molybdate or paramolybdate, and tungstate may contain more metal by weight than other derivatives, to the point that the inorganic portion exceeds the organic one, as can be seen in Fig. 5.22. These polymers have not yet been fully characterized: some data are in Tables 5.14 and 5.15 and in Fig. 5.23.

Therefore, chitosan polymolybdate can be expected to be a quite stable chromatographic support in salt solutions and in acid solutions, and to be a selective agent for the collection of phosphate because of the well-known reaction of phosphate with molybdate.[15]

Several water samples were spiked with ^{32}P phosphate and passed through columns of chitosan polymolybdate: the results are presented in Table 5.16 and Fig. 5.24.

Chitosan polymolybdate can be prepared as follows: 6·5 g of 100–200 mesh chitosan powder are suspended in 500 ml of water to which 2·5 ml of 85% acetic acid are added; the resulting pH is 3·5. In another flask, 20 g of ammonium paramolybdate tetrahydrate are dissolved in 1 l of

TABLE 5.14. COLLECTION OF METAL IONS FROM 100 ml OF SOLUTIONS, ON 5×1 cm COLUMNS OF CHITOSAN POLYMOLYBDATE, AT THE FLOW-RATE OF 3 ml min^{-1}. ALL VALUES ARE FROM TWO INDEPENDENT TECHNIQUES, MOSTLY RADIOCHEMISTRY AND POLAROGRAPHY

Ion	Concn. mM	pH	% collected
Co(II)	0·1	4·0	9
Cr(III)	0·1	4·0	22
Zn(II)	0·1	4·2	30
Cu(II)	5×10^{-3}	4·0	16
Fe(III)	0·1	4·0	20
Cd(II)	5×10^{-3}	5·0	0
Ni(II)	0·1	5·0	0
Pb(II)	5×10^{-3}	8·1	37
Hg(II)	0·4	6·1	60
Ti(III)	0·1	5·0	100
Sb(III)	0·1	2·9	100
$RuNO^{+}$	carrier-free	3·6	95
	carrier-free	6·6	99
	carrier-free	8·1 (sea-water)	81
Cs (I)	carrier-free	6·0	20

(R. A. A. Muzzarelli, original results.)

distilled water, pH = 6·5. The second solution is slowly introduced into the chitosan solution through a separatory funnel, in order to permit a complete mixing of the drops in the becker, and a large magnetic rod is then used to provide very strong stirring. The white precipitate is then filtered and washed with distilled water until the stannous chloride–thiocyanate test for molybdenum is negative. It has been suggested that it is best to lyophilize the product, or to keep it under water, to avoid the formation of a crust.

TABLE 5.15. COLLECTION OF STRONTIUM ON 10×1 cm COLUMNS OF CHITOSAN POLYMOLYBDATE OR POLYTUNGSTATE (PMC OR PTC). RADIOCHEMICAL MEASUREMENTS ON ^{90}Sr LABELLED SOLUTIONS

Column	Collection %	Elution by 10% NaCl %
Strontium chloride 0·4 mM		
PMC	100	78
PTC	100	70
Strontium nitrate 0·4 mM		
PMC	100	60
PTC	75	57
Carrier-free strontium		
PMC	100	92
PTC	100	90
Carrier-free strontium in sea-water		
PMC	50	
PTC	74	

(R. A. A. Muzzarelli, original results.)

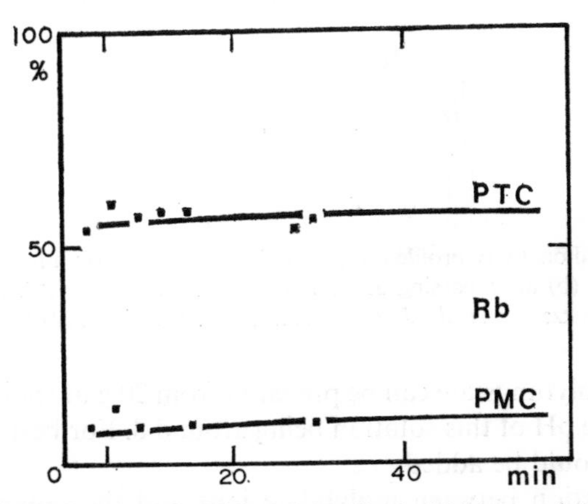

FIG. 5.23. Rates of collection of rubidium from 50 ml of 0·44 mM solution on 200 mg of chitosan polymolybdate (PMC) or chitosan polytungstate (PTC). (R. A. A. Muzzarelli, original results.)

TABLE 5.16. PER CENT COLLECTION OF 16 ng PHOSPHATE ON POLYMOLYBDATE 1×7 cm COLUMNS, FROM VARIOUS AMOUNTS OF BRINE OR WATER

Solution	Volume, ml	Collection, %
Distilled water	1000	100
Potable water	500	100
Sodium nitrate 0·5 g/l	1000	100*
Sea-water	2000	100

* See Fig. 5.24.

(From R. A. A. Muzzarelli *et al.*, *J. Radioanal. Chem.* **10,** 27 (1972).)

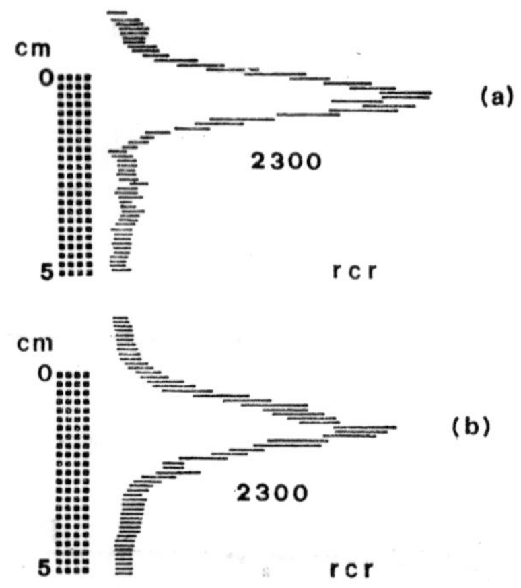

FIG. 5.24. ^{32}P radioactivity profile on a chitosan molybdate column (a) before passing sodium nitrate, (b) after passing 350 ml solution nitrate 0·5 g l^{-1}. (From R. A. A. Muzzarelli *et al.*, *J. Radioanal. Chem.* **10,** 27 (1972).)

Chitosan polytungstate can be prepared from 20 g of sodium tungstate dihydrate; the pH of this solution being around 8. For best results, some acetic acid should be added.

The interaction between molybdate ions and the polymer is highly selective: nevertheless, selective separations by the precipitation process alone are poor because, for physical reasons, other ions present in solution are carried down. For instance, most of the cobalt present together with molybdenum is found in the precipitate. Therefore, in order to make the precipitation as clean as possible and to use it as a way of performing separations, it is convenient to introduce into the mixture a complexing agent such as thiocyanate. This complexing agent may be preferred as much data exist on chromatography in thiocyanate solutions, as it can be easily handled in acidic solutions and also thiocyanate solutions are suitable for polarographic and spectrophotometric determinations. Other complexing agents such as EDTA and dithizone could be used, although they have not been tested for the present use.

The formation of chitosan polymolybdate is irreversible in the presence on thiocyanate, whose concentration in any case should be kept below 2 M to avoid any risk of precipitation of the chitosan thiocyanate.

TABLE 5.17. SEPARATIONS OF 275 mg OF MOLYBDENUM FROM OTHER IONS BY PRECIPITATION WITH CHITOSAN IN THIOCYANATE AT pH = 3·8, 40 ml FINAL VOLUME, WASHING BY 10 ml THIOCYANATE 1 N. ALL VALUES ARE FROM TWO INDEPENDENT TECHNIQUES, MOSTLY RADIOCHEMISTRY AND POLAROGRAPHY

Ion	μg metal in the Mo(VI) soln.			% Left in solution
	Initial	Found soln.	Found wash.	
Cu(II)	318	292	0	92
Pb(II)	1036	996	0	96
Ni(II)	294	228	6	80
Cd(II)	560	518	0	94
Ti(III)	240	143	50	80
Co(II)	294	268	26	100
Fe(III)	255	156	17	77
Hg(II)	1003	630	70	68
Zn(II)	326	325	0	100
Cr(III)	260	52	10	24
Mg(II)	121	121	0	100
Ca(II)	200	200	0	100
Al(III)	135	130	5	100
UO_2^{++}	1000	760	110	87
phosphate	1000	0	0	0

(R. A. A. Muzzarelli, original results).

The precipitation of chitosan polymolybdate is quantitative even in the presence of large amounts of thiocyanate.

The adopted experimental conditions were as follows: the chitosan solution in acetic acid 1 : 10 contained 20 mg ml^{-1}; the sodium molybdate solution in 1 : 10 acetic acid contained 55 mg Mo ml^{-1}. The solutions of the other ions were 1 mM. Five ml of the molybdenum solution with 5 ml of the metal ion solution, 5 ml of the thiocyanate 1 N solution, and a suitable radioisotope if required, formed a solution which was added to 25 ml of the chitosan solution, under stirring. The final volume was 40 ml and the pH was 3·8. Under these conditions thiocyanate was present in excess; a ratio around 1 : 600 existed between metal ion and molybdenum, and a large amount of chitosan was present to ensure the complete precipitation of molybdate. After stirring for 10 min, the precipitate was separated by centrifugation, collected and washed with 10 ml of thiocyanate: the aqueous phases were ready for polarographic or spectrophotometric determinations, or in most cases, for radiochemical counting. By spectrophotometry a check of the precipitation was carried out following traditional methods and it was found that only 4·02 mg out of 275·00 of molybdenum were left in solution after the said procedure, corresponding to 1·4%. The same value was obtained by polarography. Another determination was attempted by destroying the polymer in perchloric–nitric acids mixture: this yields a white precipitate where the total amount of molybdenum is present: the precipitate can be dissolved in ammonia solution and the molybdenum content can be determined by any suitable method.

The data on separations of molybdenum from other ions are reported in Table 5.17. They were obtained by both radiochemical and polarographic measurements whenever possible. The information on uranium was obtained by radiochemical techniques using ^{235}U tracer and by spectrophotometry; it is interesting to note that uranium can be separated from molybdenum and also from a large proportion of chromium, under the reported conditions.

Alkali–earth metal ions and aluminium can be separated from molybdenum even in absence of thiocyanate. The phosphate ion deserves a special mention as under these conditions it confirms the data reported for the column chromatography on chitosan polymolybdate.

The determination of molybdenum in seawater is presented in Table 5.9.

SECOND- AND THIRD-ROW POST-TRANSITION METAL IONS COLLECTION AND CHROMATOGRAPHY

The collection percentage of the second- and third-row transition and post-transition metal ions are listed in Table 5.18. They confirm the chelating ability of chitosan, particularly in the case of mercury; silver and lead are also collected to a great extent. The values for molybdate con-

TABLE 5.18. SECOND- AND THIRD-ROW TRANSITION AND POST-TRANSITION METAL ION COLLECTION ON 200 mg CHITIN AND CHITOSAN. PER CENT OF THE AMOUNT PRESENT IN 50 ml OF 0·44 mM AQUEOUS SOLUTION (ATOMIC ABSORPTION SPECTROMETRY)

	pH		hr	Mo(VI)	Ag(I)	Sn(II)	Sb tartrate	Hg(II)	Pb(II)
Chitin	2·5		1	70	20	52	88	12	5
			12						
		EDTA	1		11		4	11	4
	5·5		1	0	41			42	36
			12		70			55	
		EDTA	1					29	2
Chitosan	2·5		1	100	80	17	38	98	26
			12	100	96	96		100	48
		EDTA	1		43		16	58	13
	5·5		1	0	100			100	95
			12					100	99
		EDTA	1					89	2

(R. A. A. Muzzarelli *et al.*, *Talanta* **16,** 1571 (1969); see also U.S. Patent 3,635,818 (1972).)

firm the findings relevant to the precipitation of chitosan polymolybdate, while the absence of collection of molybdate at pH = 5·5 indicates that molybdate can be eluted from a chromatographic chitosan column by simply rising the pH, with sodium carbonate, for instance. Antimony was used in form of tartrate; nevertheless, it is collected very satisfactory by chitin. 0·1 M EDTA can prevent the collection of antimony on chitin in acidic media, but its presence does not lower appreciably the values for

TABLE 5.19. CHROMATOGRAPHIC RESULTS FOR TRACE METAL SEPARATIONS FROM 38 AND 1 mM SOLUTIONS OF THALLIUM(I) NITRATE (100 ml ON A 15×1 cm 100–200 MESH CHITOSAN COLUMN. ELUENT VOLUME 100 ml)

Ion	Eluent	% Recovery from 38 mM	% Recovery from 1 mM
Silver	NH_4OH+NH_4Cl	100	100
Cadmium	KCN 0·1 N	100	100
Indium	EDTA 0·1 M		95
Mercury	KCN 0·1 M	94	99
Lead	NH_4COOCH_3 2 M	100	100
	KCN 0·1 M	0	0
Nickel	EDTA 0·1 M	100	100
Terbium	EDTA 0·01 M	90	90

(From R. A. A. Muzzarelli. *et al. Mikroch. Acta* 892 (1970).)

mercury on chitosan: in fact, it is impossible to elute mercury from chitosan columns, and KCN is required.

Among the elements, silver, cadmium, indium, tin, antimony, mercury, thallium, lead, and bismuth, only thallium forms very scarce complexes, and therefore it was anticipated that a separation of thallium from other elements would be feasible on chitosan.[6] In practice, the isolation of trace amounts of these metals from large amounts of thallium matrix are demanded in various fields.

For the experiments which have been carried out, thallium(I) nitrate solutions were 1 and 38 mM; the other metal ion solutions were 1 mM; the data were obtained by both radioisotope techniques and polarography. For a first series of experiments, 100 ml of the 38 mM thallium solution was submitted to chromatography as follows: after the passage of the original 100 ml of solution through the column (1×15 cm, 100–200 mesh powder), 50 ml of water were used for washing. Elutions of the other ions were performed as indicated in Table 5.19. Generally speaking, the thallium recovery was 98% in about 150 ml solution, and the trace metal recovery after elution was between 90 and 100%.

A second series of experiments was carried out with a 1 mM solution of thallium and other ions. The results are shown in Table 5.19. At this concentration a sensible fraction of the thallium is left on the column because the recovery is 78–87%. Recoveries of the other elements after selective elution are higher than 95%: these reagents do not elute the remaining fraction of thallium. Retention of thallium reportedly depends on temperature and does not seem to be in competition with the uptake of other ions: its recovery could reach 98% at 80 °C. The other ions are retained in the upper part of the column, as shown by radiochromatographic scanning. Fe(III) gives a coloured band which is visible at the

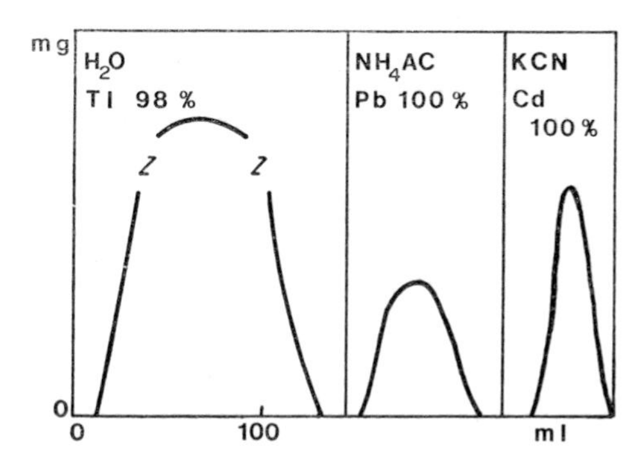

Fig. 5.25. Purification of thallium (I) nitrate by chromatography on a 15×1 cm column of chitosan, 100–200 mesh, 25 °C. 100 ml of 38 mM Tl and 1 mM Cd and Pb solution passed at a flow rate of 300 ml/hr. Selective elutions performed with 2M ammonium acetate and 0·1 M potassium cyanide for Pb and Cd respectively. (From R. A. A. Muzzarelli *et al.*, *Mikroch. Acta* 892 (1970).)

top of the column after passage of a solution containing 1 g of thallium nitrate (100 ml of 38 mM solution, originally containing 40 μg of Fe(III). The elution curves of lead and cadmium are given in Fig. 5.25.

The results for mercury, including the synthetic chelating polymer Dowex A-100 are also presented in Fig. 5.26, for various concentrations of metal ion: apparently the affinity of mercury for these polymers is very high.[3]

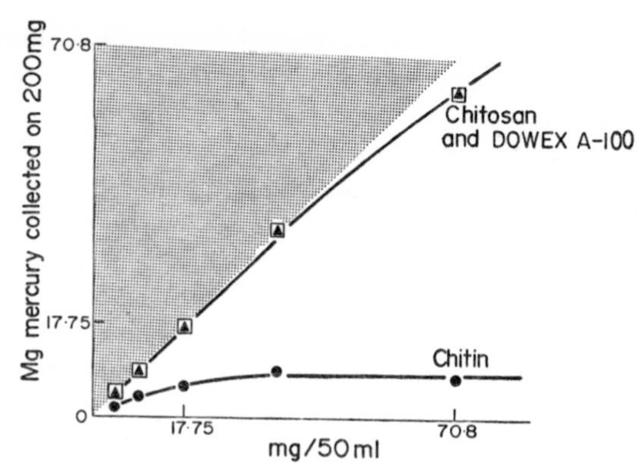

FIG. 5.26. Collection of mercury on chitosan, Dowex A-100 and chitin. (From R. A. A. Muzzarelli *et al., J. Chromatogr.* **47,** 414 (1970).)

Data on the collection of methyl-203-mercury acetate and 203-mercury(II) acetate from "waste" solutions are reported in Table 5.20. This data confirms that the most efficient substituted cellulose is unable to collect mercury in these forms, while chitin, chitosan and Dowex A-100 can collect it. The distribution coefficients for metallorganic mercury are particularly good for chitosan and Dowex A-100.[16]

TABLE 5.20. COLLECTION DATA FOR 30 mg METHYL-203-MERCURY ACETATE OR 30 mg MERCURY(II) ACETATE IN 50 ml OF ACETIC ACID + ACETALDEHYDE SOLUTION, pH = 4·8 AT 18 °C, ON 200 mg POLYMER 100–200 MESH

Polymer	Per cent collection				Distribution coefficients, K_d	
	1 hr		24 hr			
	m.o.	ionic	m.o.	ionic	m.o.	ionic
DE Cellulose	n m	n m	n m	n m	—	—
Chitin	26	22	29	22	113	76
Chitosan	50	65	47	92	244	2790
Dowex A-100	79	80	89	86	2283	1730

(From R. A. A. Muzzarelli *et al., Water, Air and Soil Pollution* **1,** 65 (1971).)

In fact, when 16 cm columns are operated, 92% of the 30 mg methyl-203-mercury acetate is collected on chitin, from a 50 ml solution passed, while on chitosan and Dowex A-100 columns under similar conditions the total amount is collected.

TABLE 5.21. PER CENT COLLECTION OF 30 mg METHYL-203-MERCURY ACETATE ON 1 cm i.d. CHROMATOGRAPHIC COLUMNS, FLOW-RATE 3 ml/min, TEMPERATURE 18 °C

Water (ml)	Washing (ml)	Column height (cm)	Per cent collection		
			chitin	chitosan	Dowex A-100
Demineralized water, pH = 6·0					
50	50	16	83	94	100
50	50	24	95	100	100
5000	—	24	21	45	88
Drinking water, pH = 7·5					
2000	—	16	—	100	84
Acetic acid+acetaldehyde, pH = 4·8					
50	50	16	92	100	100
2000	—	16	—	61	100
Sea-water, pH = 8·1					
50	50	16	20	81	85
1000	—	24	3	22	40

(From R. A. A. Muzzarelli *et al.*, *Water, Air and Soil Pollution* **1**, 65 (1971).)

In the chitin column the chromatographic band is very broad and begins to move out with the 50 ml washing solutions; in the chitosan and Dowex A-100 most of the mercury is present in the very first centimetre of the column. Tailing in the chitosan columns is a little bit more evident than in Dowex A-100 column.

The results for various waters polluted with methyl-203-mercury acetate are reported in Table 5.21.

One can observe that the performance of Dowex A-100 and chitosan are quite good under these conditions. In view of its rather high affinity for magnesium, Dowex A-100 does not perform so well as chitosan in drinking water, while in other cases it gives the impression of working slightly better, which is also due to the fact that in a 16 cm column there is much more Dowex A-100 than chitosan by weight. In any case, one should keep in mind that mercury is the most effectively fixed element on Dowex A-100.[15] On the other hand, chitosan does not swell, while swelling is most evident in Dowex A-100.

Temperature plays a certain rôle in collection of metal ions on polymers, and, as far as chitin and chitosan are concerned, collection is generally better at low temperatures. Trials at 2 °C confirm this trend in the case of methyl mercury acetate. Fifty-three per cent of 30 mg in 4000 ml were fixed on a 24 cm chitosan column, and 100% of the same amount in 50 ml were fixed. Of course, the dilution affects collection yields and below a certain threshold fixation, is not completed, even with a long column. Elution of methyl mercury acetate can therefore be performed by countercurrent washing with hot water at temperatures near 100 °C from a jacketed column kept at 100 °C. Recovery of methyl mercury acetate with 40 ml water was 60% in the first 20 ml fraction, and 30% in the second, with traces in the third.

ALUMINIUM AND LANTHANONS

The rates of collection of aluminium, scandium, and lanthanum have been measured on 0·44 mM solutions, 50 ml of which were stirred with 200 mg of polymers. The quantitative determinations were done, for the elements mentioned, by u.v. spectrophotometry after treating the filtered solutions with oxime.

Aluminium at pH 4·0 is collected to the extent of 80% after 6 min, but after 40 min only 20% of the aluminium present is held by chitosan: apparently, as aluminium initially interacts with a basic insoluble substance, it can be locally hydrolysed, but a few minutes later the equilibrium is established again, and it seems therefore that under these conditions aluminium is slightly collected by chitosan. At higher pH, for instance 6·2 aluminium is quantitatively retained by the polymer, but this is due to hydrolysis.

Scandium is collected on both polymers at pH 5·5 under the conditions described above, to the extent of 85% after 4 min and to 92% after 1 hr.

At the same pH 5·5 lanthanum after 1 hr is collected to the extent of 15% on chitin and 20% on chitosan.

Lanthanides as well as lanthanum generally show slower rates and both chitin and chitosan have limited capacity for them: data for Ce(III)

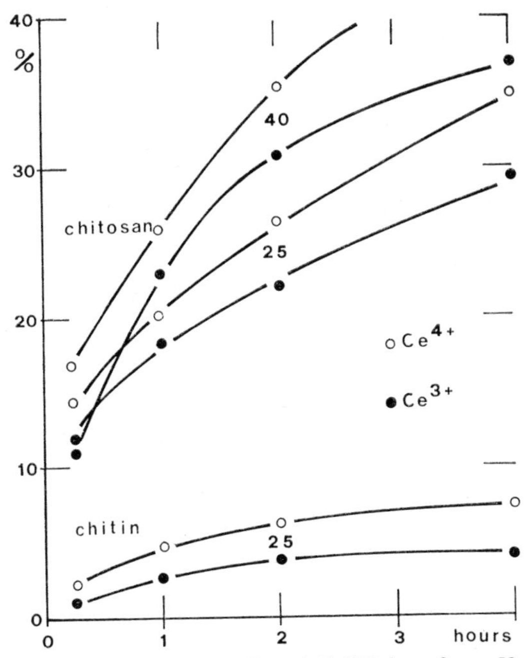

FIG. 5.27. Rates of collection of Ce(III) and Ce(IV) ions from 50 ml of 1·76 mM solutions at 25° and 40 °C, pH 6·0, on 200 mg chitin and chitosan, 100–200 mesh. (From R. A. A. Muzzarelli *et al.*, *J. Radioanal. Chem.* **10**, 17 (1972).)

and Ce(IV) are in Fig. 5.27 while data for terbium, europium and thulium are listed in Table 5.22. Generally speaking, lanthanides are less effectively collected than transition elements on both polymers.

Further investigations at lower pH, might possibly lead to a separation procedure of scandium from other group III A and B elements.

TABLE 5.22. PER CENT COLLECTION OF SEVERAL LANTHANIDES FROM 1·76 mM SOLUTIONS AT pH = 6·0, 50 ml WITH 200 mg POWDER, 100–200 MESH

Contact time, min	Chitin		Chitosan	
	25 °C	40 °C	25 °C	40 °C
$^{152,154}Eu$				
15	5	3	30	26
60	7	5	33	30
120	8	5	37	34
240	9	6	45	38
^{170}Tm				
15	3	0	31	40
60	4	0	35	42
120	4	0	37	42
240	5	0	38	43
^{160}Tb				
15	9		32	
60	6		36	
120	4		39	
240	4		43	

For Ce(III) and Ce(IV) see Fig. 5.27.

(From R. A. A. Muzzarelli *et al.*, *J. Radioanal. Chem.* **10,** 17 (1972).)

SPECIAL CHROMATOGRAPHIC APPLICATIONS OF CHITIN AND CHITOSAN

Chromatographic separations of organic substances on chitin or chitosan columns have been reported in two papers: lysozymes from hen egg white and from turnip were isolated by column chromatography on chitosan. Elution was accomplished with acetic acid solutions.

Chitin columns, 17×1·3 cm equilibrated with hydrochloric acid solutions at pH 6·8 were used in chromatography of the tobacco mosaic virus. 0·5 ml of tobacco mosaic virus solution containing 0·3–2·0 mg of virus in 0·01 M sodium phosphate buffer, pH 6·8 was added to the column and the virus washed into it with three 1-ml aliquots of hydrochloric acid. The column was then connected for gradient elution with progression from water to 0·5 M K_2HPO_4. The eluate was collected at 10 min intervals in 4·3 ml aliquots, and monitored for optical density at 254 nm.

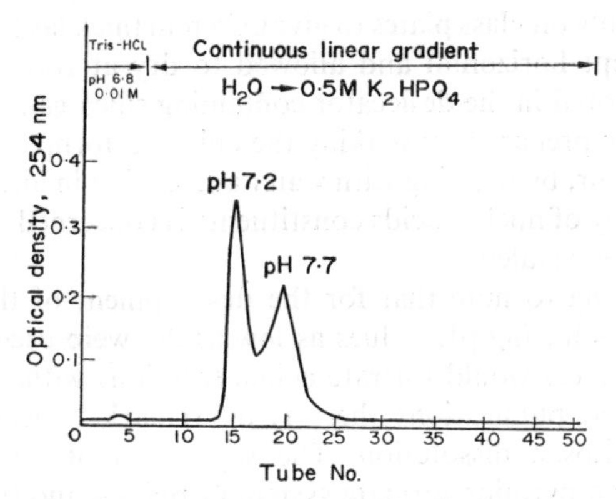

FIG. 5.28. An elution curve of tobacco mosaic virus from a chitin chromatographic column. (From P. M. Townsley, *Nature, Lond.* **191**, 626 (1961).)

Results, as presented in Fig. 5.28, were reproducible, but the amount of virus in the two peaks varied. Material from both the peaks was infective and birefringent. The material eluted at pH 7·2 was four to ten times as infective as that eluted at pH 7·7.

Chitin was found to be very useful for this separation as the treatment for adsorption and elution is very mild. That author anticipated that chitin can be used for nucleic acid chromatography. However, 10 years had to pass before reading results concerning nucleic acids on chitosan. In the meantime, thin layer chromatography of nucleic acid constituents has been widely accomplished on diethylaminoethyl, ecteola, and polyethyleneamine celluloses, on which clean and rapid separations of minute amounts of these constituents were possible.

Chitosan can be conceived, in this light, as a new type of anion-exchange material, and two papers dealing with ion-exchange chromatography of nucleic acid constituents on chitosan-impregnated cellulose thin layers, have been published.

Powdered chitosan (2·4 g), prepared according to the method of Wolfrom were dissolved in 60 ml of 2·5% (w/v) formic acid and diluted to 300 ml with distilled water, and then filtered on a glass filter. A suspension of microcrystalline cellulose (Avicel SF) made with 15 g in 60 ml of the resulting 0·8% w/v chitosan solution was homogenized in a glass homogenizer for 25 sec. After deaeration with suction, the suspension was spread evenly on glass plates to give 0·25 mm thick layers. The coated plates were kept horizontal and allowed to dry at room temperature before being stored in the desiccator containing silica gel. Layers of free chitosan can be prepared by soaking the chitosan formate layers in 1% ammonia for 1 hr, by washing with water and drying in air. As far as the chromatography of nucleic acids constituents is concerned, the two types of layers were equivalent.

It is interesting to note that for the development of these chitosan layers, solutions having pH values as low as 2·5 were used; this means that chitosan layers would tolerate acidic solutions without dissolving, and in the cited publications there is no remark indicating troubles because of chitosan dissolution. The separation of nucleotides was achieved using a pyridine formate system at pH 4·4 and 0·25 M ammo-

TABLE 5.23. R_f VALUES FOR NUCLEOSIDES IN VARIOUS SOLVENTS

Solvent: A = water; B = methanol–water (1:1); C = ethanol–water (4 : 1); D = 0·05 M formic acid, pH 2·5; E = 0·5 M pyridine, pH 9·2; F = 0·5 M triethylammonium carbonate, pH 7·3

Compound	Solvent					
	A	B	C	D	E	F
Inosine	0·60	0·45	0·40	0·64	0·63	0·74
Adenosine	0·42	0·39	0·51	0·43	0·51	0·44
Guanosine	0·45	0·37	0·31	0·46	0·54	0·63
Cytidine	0·63	0·56	0·47	0·71	0·70	0·69
Uridine	0·71	0·57	0·60	0·75	0·72	0·77
Deoxyadenosine	0·40	0·44	0·63	0·43	0·51	0·45
Deoxyguanosine	0·46	0·43	0·50	0·48	0·56	0·64
Deoxycytidine	0·63	0·62	0·67	0·74	0·69	0·69
Deoxyuridine	0·70	0·61	0·61	0·77	0·70	0·77
Thymidine	0·70	0·66	0·66	0·79	0·72	0·72

(From Nagasawa *et al.*, *J. Chromatogr.* **47,** 408 (1970).)

TABLE 5.24. R_f VALUES FOR NUCLEIC BASES IN VARIOUS SOLVENTS
Solvent: A = water; B = methanol–water (1 : 1); C = ethanol–water (4 : 1); D = 0·05 M formic acid, pH 2·5; E = 0·5 M pyridine, pH 9·2; F = 0·5 M triethylammonium carbonate, pH 7·3

Compound	Solvent A	B	C	D	E	F
Adenine	0·28	0·46	0·73	0·29	0·32	0·33
Hypoxanthine	0·42	0·55	0·62	0·50	0·50	0·58
Guanine	0·21	0·33	0·39	0·23	0·29	0·21
Cytosine	0·49	0·64	0·67	0·70	0·64	0·65
Uracil	0·58	0·66	0·80	0·70	0·63	0·67
Thymine	0·58	0·71	0·92	0·70	0·65	0·63

(From K. Nagasawa *et al.*, *J. Chromatogr.* **47**, 408 (1970).)

nium formate at pH 4·0 and 6·5, and other solutions including 0·05 M formic acid, at pH 2·5. Results are reported in Tables 5.23 and 5.24. The role of chitosan on the layers is quite evident from Fig. 5.29 where a comparison of microcrystalline cellulose alone is made with the same

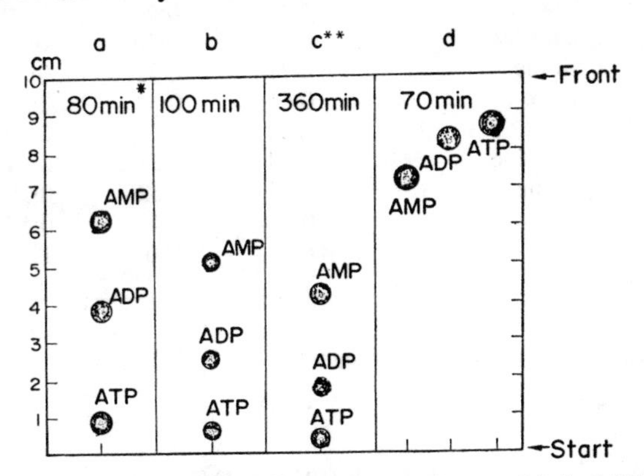

FIG. 5.29. Influence of the chitosan formate content of layers of Avicel SF on the separation of nucleotides. The nucleotides (each 1 μm mole) were chromatographed on layers (0·25 mm thick) prepared with 0·4% (a) 0·8% (b) and 1·6% (c) chitosan formate solutions and 0·25 M pyridine formate buffer (pH 4·4). (d) Avicel SF thin layer (0·25 mm thick). * Development time. ** In this case, a severely irregular solvent front was formed, so that the R_f values could not be calculated. (From K. Nagasawa *et al.*, *J. Chromatogr.* **47**, 408 (1970).)

impregnated with different amounts of chitosan, but a too large amount of chitosan brings about irregularities of the front and long development times.

For the separation of nucleotides, 0·25 M pyridine-formate at pH 2·6 was the best solvent system, as from data in Table 5.25. Also 0·1 M formic acid was most suitable for separation within a reasonable development time.

TABLE 5.25. R_f VALUES FOR DINUCLEOSIDE-3′ → 5′ PHOSPHATES IN WATER OR PYRIDINE

Formic acid systems		Pyridine–formic acid pH			Water-formic
Compound		2·6	3·0	4·4	0·2 M
3′-adenylyl est.	adenosine	0·61	0·45	0·33	0·67
	guanosine	0·37	0·30	0·41	0·39
	cytidine	0·61	0·56	0·44	0·76
	uridine	0·47	0·41	0·43	0·41
3′-guanylyl est.	adenosine	0·39	0·34	0·36	0·39
	guanosine	0·16*	0·32*	0·32*	0·10*
	cytidine	0·52	0·48	0·41	0·50
	uridine	0·27	0·26	0·38	0·10
3′-cytidylyl est.	adenosine	0·63	0·63	0·50	0·73
	guanosine	0·53	0·59	0·47	0·52
	cytidine	0·71	0·71	0·57	0·76
	uridine	0·66	0·58	0·53	0·56
3′-uridylyl est.	adenosine	0·54	0·52	0·49	0·49
	guanosine	0·24	0·28	0·44	0·11
	cytidine	0·66	0·62	0·58	0·58
	uridine	0·38	0·43	0·57	0·13
Develop. min		80	100	120	400

* Tailing.

(From K. Nagasawa *et al.*, *J. Chromatogr.* **56**, 378 (1971).)

REFERENCES

1. R. A. A. MUZZARELLI and O. TUBERTINI *Talanta* **16**, 1571 (1969); see also U.S. Patent 3,635,818 (1972).
2. R. A. A. MUZZARELLI and M. MARINELLI, *Inquinamento* **14**(3), 29 (1972).
3. R. A. A. MUZZARELLI, G. RAITH and O. TUBERTINI, *J. Chromatogr*. **47**, 414 (1970).
4. R. A. A. MUZZARELLI, *Anal. Chim. Acta* **54**, 133 (1971).

5. M. VIAL, J. P. QUILES, M. YANAKIEVA and M. CHENE, *Cellul. Chem. Technol.* **3,** 21 (1969).
6. R. A. A. MUZZARELLI and O. TUBERTINI, *Mikroch. Acta* 892 (1970).
7. R. A. A. MUZZARELLI and L. SIPOS, *Talanta* **18,** 853 (1971).
8. R. A. A. MUZZARELLI, *Rév. Int. Océanogr. Médicale* **21,** 93 (1971).
9. W. L. HAMITER and J. M. COLLINS, U.S. Patent 3,108,897 (1963).
10. S. ADACHI, Japan Patent 7,002,799 (1970).
11. R. A. A. MUZZARELLI, R. ROCCHETTI and G. MARANGIO, *J. Radioanal. Chem.* **10,** 17 (1972).
12. J. E. KIEFER and G. P. TONEY, *Ind. Eng. Chem. PRD* **4,** 253 (1965).
13. V. B. ALESKOWSKII, T. I. KALININA and N. A. TYUTINA, *Zh. Prikl. Khim.* **40,** 1279 (1967).
14. H. MIMOUN, I. SEREE DE ROCH and L. SAJUS, *Bull. Chem. Soc. France* 1937 (1972).
15. R. A. A. MUZZARELLI and B. SPALLA, *J. Radioanal. Chem.* **10,** 27 (1972).
16. R. A. A. MUZZARELLI and A. ISOLATI, *Water, Air and Soil Pollution* **1,** 65 (1971).
17. R. A. A. MUZZARELLI and M. MARINELLI, *Inquinamento* **14(4)**, 27 (1972).
18. P. M. TOWNSLEY, *Nature, Lond.* **191,** 626 (1961).
19. K. NAGASAWA, H. WATANABE and A. OGANO, *J. Chromatogr.* **47,** 408 (1970
20. K. NAGASAWA, H. WATANABE and A. OGANO, *J. Chromatogr.* **56,** 378 (1971).
21. TAKEDA and T, TOMIDA, *J. Shimonoseki Univ. Fish.* **18,** 36 (1969).
22. TAKEDA and T. TOMIDA, *J. Shimonoseki Univ. Fish.* **20,** 107 (1972).

CHAPTER 6

OTHER CHELATING POLYMERS

THE natural chelating polymers described in the previous chapters are those which, at the present time, have caught the attention of the analytical chemists and have found applications in the field of chromatography and of trace metal separation.

In addition, there are many other natural polymers exhibiting chelating ability, but they have been scarcely studied from this standpoint, as they are sometimes difficult to isolate, or occur in small amounts, or because their formulae are unknown. In this concluding chapter some of these will be recalled, in order to underline their important chelating ability, which is often forgotten because of the scant information available.

POLYSACCHARIDE DERIVATIVES

Table 6.1 shows the selectivity coefficients for a number of natural chelating polymers, obtained for equimolar amounts of the two cations in solution, under comparable conditions.[1] The results indicate that a high selectivity for copper compared to calcium, is a characteristic feature of polyanions containing carboxyl groups in general. It was also remarked that a positive enthalpy change accompanies the reactions involving copper with alkaline earth ions: this is to indicate that the driving force in this group of exchange reactions must be a favourable entropic change.

A characteristic of these data seems to be the lack of importance of the detailed structure of the polyanion, and particularly significant is the

TABLE 6.1. SELECTIVITY COEFFICIENTS OF ANIONIC POLYMERS.*

Anionic group	Polymer	k_{Mg}^{Ca}	k_{Mg}^{Sr}	k_{Mg}^{Ba}	k_{Ca}^{Co}	k_{Ca}^{Cu}
—COOH	Polyguluronate	40	150	130	0·17	8
	Polymannuronate	1·8	1·2	10	1·1	12
	Alginate, 92% mannuronic acid	1·6	1·4	13	1·0	11
	Acetylated polyguluronate	1·0	1·0		1·2	12·0
	Pectate	7·0	9·6	10·1	0·7	32
	Oxidized cellulose	1·1	1·5	3·5	1·1	11
	Carboxymethyl dextran	1·8	1·4	1·3	1·2	10
	Carboxymethyl cellulose	1·4	1·3	1·6	1·3	17
	Polyacrylate	1·1	0·7	3·0	1·7	25
	Hyaluronate	0·9				3·0
—COOH and —SO_3H	Chondroitin sulphate	1·1	0·95	1·3		2·4
	Ascophyllan	1·1	1·3	2·1	0·9	7·0
—SO_3H	Fucoidan	1·3	1·6	2·9	1·0	1·5
	Dextransulphate	1·1	2·0	7·0	1·2	0·9
	Carrageenan	1·2	1·4	1·6	1·2	1·2
—$OP(OH)_2O$	Polyphosphate	0·9	1·1	6·0	1·2	8·0

* The values for polyguluronate and polymannuronate are determined by extrapolations, those for pectate correspond to a cationic composition of approximately 50% of each of the two ions bound to the pectate, while the rest of the figures are determined with equivalent amounts of the two ions in solution

(From A. Haug *et al.*, *Acta Chem. Scand.* **24,** 843 (1970).)

observation that this selectivity is not destroyed by acetylation. This implies that the binding mechanism which is responsible for the higher affinity of copper relative to calcium ions in carboxylate-containing polyanions is independent of the presence of hydroxyl groups in the polymer and of the steric arrangements of the carboxyl groups. The low copper–calcium selectivity of chondroitin sulphate and hyaluronate might indicate a lower affinity of copper compared to calcium for polysaccharides containing acetamido groups, but the data relevant to chitin do not allow this hypothesis.

The selectivities of the polyuronides in exchange reactions between alkaline earth ions depend very much upon the detailed structure of the polymer, as any selectivity is completely lost upon acetylation of the hydroxyl groups.

However, when the polyuronide allows a preferential binding of calcium compared to magnesium, copper is probably bound prefentially by the same mechanism; in addition, the mechanism making the binding of copper to carboxyl groups energetically more favourable than the binding of calcium is operative.

In order to assess the mechanism of interaction, the dimensions of the molecules should be known, but this is not yet the case.

Polygalacturonic acid obtained from pectin forms sparingly soluble metal complexes;[2] the maximum amount of copper ions complexed by polygalacturonic acid was obtained with a ratio of 0·7 of meq Cu/l to meq unit polymer/l. Increasing amounts of polygalacturonic acid had a negligible effect. The same ratio was observed for zinc, cadmium, and nickel. For these elements, the per cent metal ion collected was higher than 75% at pH 4·5, as reported in Fig. 6.1. If all of these ions are present in the same solution, the collection decreases in the order: Cu, Cd, Zn, and Ni. This polymer can be used repeatedly for collection of transition metal ions by complexation. After each collection stage, the polymer is regenerated by hydrochloric acid. The loss of polymer after the first regeneration treatment was 5% w/w with 1 N HCl. Lower losses were recorded for subsequent cycles, and for other transition elements. Ag(I), Ca(II), Ce(III), Mn(II), Fe(II), and Fe(III) can be precipitated by polygalacturonic acid as complexes. Metals which are complexed in anionic forms such as Au, Pt, Ru, Rh, Cr, and Pd can be precipitated from solutions by adding first polyethyleneimine followed by polygalacturonic acid, "sandwiching" the anion between the two polyelectrolytes.

Metal ions complexed in solutions by a polyanion precipitate as insoluble "sandwiched" complexes on addition of a polybase. While this has been done with synthetic polyacids and polybases like polymetacrylic acid and polyethyleneimine, it seems that this has not yet been tried with natural polymers.

Studies on the strength of calcium binding to the carboxyl groups of pectin were carried out by Kohn.[3-7] The interactions of strontium, calcium, and barium with carboxyl groups of pectate were defined on the basis of the selectivity coefficients for exchange of calcium for strontium and calcium for barium in pectate, and from the stability constants of calcium, strontium, and barium pectates in KCl solutions. The affinity is

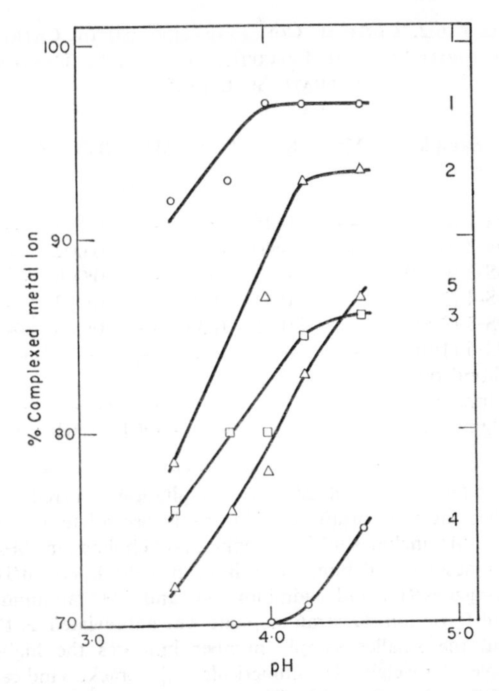

FIG. 6.1. Weight percentage of metal ions collected as a function of pH. Initial conceutrations in meq/l or unit meq/l, respectively: (1) Cu, 4·05 and PGUA 5·68; (2) Cd 2·52 and PGUA 4·12; (3) Zn, 5·00 and PGUA 6·67; (4) Ni 5·00 and PGUA 5·68; (5) Cr 3·32 and PGUA 6·39. (From H. H. G. Jellinek *et al.*, *Water Res.* **6**, 305 (1972).)

larger for barium, in accordance with the data by Haug.[1] However, in the view of Kohn, calcium ions are bound first of all by electrostatic forces.

A 0·001% solution of partially esterified pectin was used to clarify the colloidal solutions resulting from the sulphuric acid digestion of titanium-bearing materials. After 4 hr at 55–60 °C complete settling of the solid occurred; clarification by this method leaves only 4% suspended solids.[8] For medical applications, polyester fibres were heparinized and could maintain a high level of heparin in contact with NaCl solutions.[9]

In order to study the selective accumulation of potassium in the living cell, due to chondroitin sulphate, which is known as the polyelectrolyte carrying ionizable sulphate and carboxyl groups which constitute the connective tissue, sodium dextran sulphate and sodium alginate were

TABLE 6.2. CRITICAL CONCENTRATION (M) OF CATIONS REQUIRED FOR THE PRECIPITATION OF 1 % DEXTRAN SULPHATE SOLUTION*

Sample	M^+	K^+	Cs^+	M^{2+}	Ba^{2+}	M^{3+}
DS-1 (2·0)	—	0·35	0·28	—	0·015	—
DS-2 (0·2)	—	0·40	—	—	0·002	—
DS-3 (0·75)	—	—	—	—	0·006	—
DS-4 (2·0)	—	0·43	0·69	—	0·012	—
DS-5 (2·5)	—	0·26	0·39	—	0·016	—
DS-6 (1·0)	—	—	—	—	—	—
Chondroitin sulphate (0·5)	—	—	—	—	—	—
Alginate (0)	—	—	—	<0·01	<0·01	<0·01

* The sign — indicates no precipitation occurred even when the concentrations of inorganic electrolyte reached 2 M. M^+ includes lithium, sodium and choline ion, M^{2+}, magnesium, calcium, strontium, nickel(II), cobalt(II), manganese(II) and cadmium ion, and M^{3+} aluminum and ferric ion. Dextran sulphate was abbreviated as DS and the smaller sample number indicates the higher molecular weight. The numericals in the brackets indicate the average number of sulphate ion per hexose ring.

(From H. Kimizuka *et al.*, *Bull. Chem. Soc. Japan* **40**, 1281 (1967).)

taken respectively as species carrying one of these groups. In Table 6.2 it may be seen that K, Cs, and Ba show a specific interaction with dextran sulphate, while the other ions studied, among which were some transition metal ions, had no specificity. It is likely that the formation of a precipitate may be attributed to their smaller hydrated ionic radii, to the nature of their electron shells and to the resulting conformation of the salts. The specific interaction of potassium with dextran sulphate is evident in Table 6.3.[10]

Chondroitin sulphate showed no specific interaction with any of the ions examined. These results indicate that chondroitin sulphate scarcely interacts with transition metal ions, and that the polymeric structure of the hexose sulphate ring is responsible for the specific interaction with potassium in the connective tissue.

Hyaluronic acid

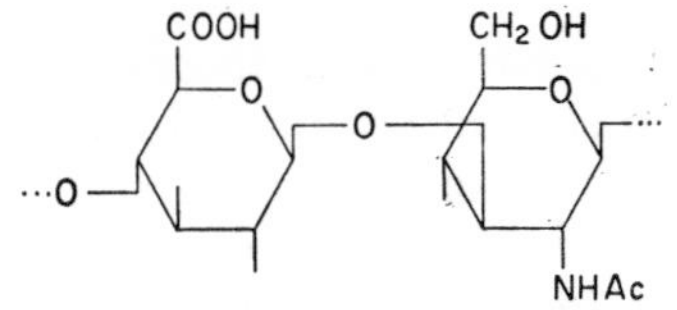

Chondroitin sulphate

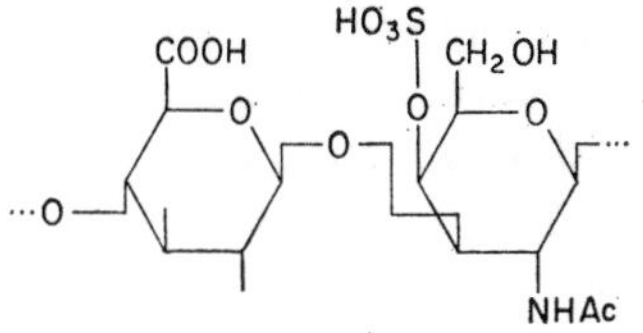

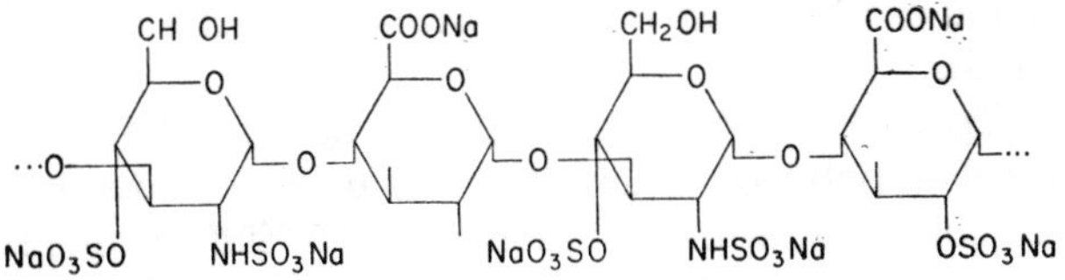

Heparin

TABLE 6.3. CRITICAL CONCENTRATION OF CATIONS REQUIRED FOR THE PRECIPITATION OF 12% DEXTRAN SULPHATE SOLUTION*

Added electrolyte	Critical concn. (M)
NaCl	6·0
KCl	0·28
RbCl	0·52
CsCl	0·44
$BaCl_2$	0·026

* The intrinsic viscosity of the dextran sulphate (DS-5) in 1 M sodium chloride solution was 0·028. The incipient aggregation of the dextran sulphate did not occur with the addition of saturated lithium chloride solution and in the presence of the 2 M solutions of the other ions described in Table 6.2.

(From H. Kimizuka *et al.*, *Bull. Chem. Soc. Japan* **40,** 1281 (1967).)

Results in agreement with those cited above were obtained by other workers, who studied dextran sulphate, chondroitin sulphate, alginate, and heparin. The change in the aggregation state produced by the addition of K, Rb, Cs, Li, Na, Ba, Sr, Ca, Mg, Ni, Mn, Co, Cd, Fe(III), and Al chlorides was interpreted as a measure of ion binding. High sulphate content and high molecular weight of the polymers contributed to aggregation, while the carboxyl groups removed the selectivity strongly.[11]

Polysaccharides are present in the soil humus.[12] With respect to quantity, they constitute the second most important component of the soil organic fraction. The polysaccharide fraction appears to be active in soil aggregate formation or binding of soil particles and probably is important in chelation and exchange reactions involving active constituents such as the uronic acids.

The occurrence of the polysaccharides fraction indicates that certain plant or microbial polysaccharides or their partial degradation products from salts or complexes with metal ions, clays or humic complexes, are resistant to decomposition. Individual polysaccharides like starch, mannan, pectin substances, xylan, and others may be almost entirely decomposed in a few weeks. Cellulose decomposes somewhat less quickly, while chitin is still more slowly utilized.[13, 14]

Studies with uronic acid containing microbial and plant polysaccharides have indicated that salt or complex formation with certain metal cations may markedly influence the rate of decomposition of the polymers in soils. In general, Al exerted the least effect and Cu the greatest, while Zn and Fe were intermediate. Some examples are presented in Fig. 6.2. The complexing of polysaccharides with metal ions may reduce the ability of microbial enzymes to hydrolyse the polymers and, under natural conditions, complexation is probably one factor responsible for the presence of a relatively stable polysaccharide fraction in the soil.

The binding activity of the polysaccharides is related to their length and structure, which allows them to bridge the space between the soil particles, to the large number of hydroxyl groups which may be involved in hydrogen bonding, and to functional groups which may allow ionic binding through di- and trivalent cations to ion-exchange sites in the clays or anion adsorption to positive charge sites on clay edges.

Most of the principal chelating groups are present in soil organic matter. Amino, imino, keto, hydroxy, thioether, carboxylate, and phosphonate groups are present in compounds which have been isolated from

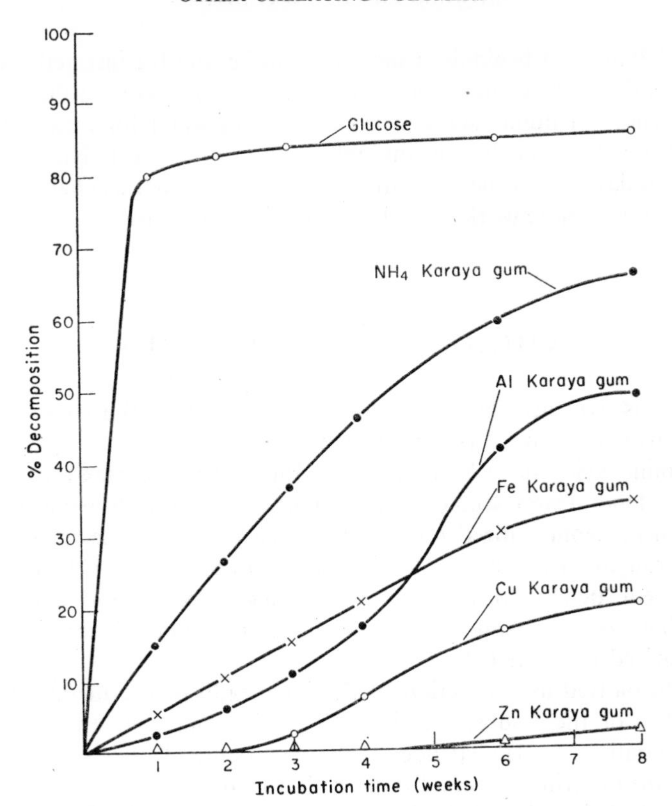

FIG. 6.2. Decomposition of karaya gum and its Fe, Al, Cu salts or complexes in Greenfield sandy loam. For comparative purposes the decomposition rate of glucose is indicated. 0·5 g organic material was added to 100 g soil portions. (From J. P. Martin, *Soil Biol. Biochem.* **3,** 33 (1971).)

soil organic matter. Ligand groups on polymers may be found in arrays which are sterically favourable for the chelation of particular metal ions.[15–17]

However, only very recently preliminary data were published on crystallinity of humic acids,[18] and therefore at the present time it is impossible to describe in quantitative chemical terms the interactions of the metal ions with humus. The molecular weight was found to have a minimum value of 1393, and a particle weight of 26,700 was calculated for humic acids suspended in water. Gas chromatography and mass spectrometry studies[19] are just beginning; the same can be said for infrared spectrometry in the study of humic and fulvic acids.[20] A study on the possible interferences of humic acids in the analysis of water for iron has been published.[21] In conclusion, our information on these polymers is too limited today, and while all authors state that they act as chelating substances, much more work is needed to obtain a clear insight.

AMINO ACID DERIVATIVES

There is evidence that metal ions are involved in the structure and function of collagen. Calf-skin collagen,[22] egg-shell matrix,[23] gelatine[24] and amino acids have been studied. Human tibial cortex and human Achilles tendon were studied in two kinds of experiments:[25] in the first experiment demineralized bone and washed tendon in the same beaker were brought in contact with solutions of one metal ion. In a second experiment, demineralized bone and demineralized tendon were separately immersed in solutions of several ions together. The results are summarized in Table 6.4.

When plotted as a function of the ionic radii of the metal ions, the capacities of these substrates identified two groups of ions which are preferentially bound independently of the specific identity of the charge and of the experimental conditions, as in Fig. 6.3.

A general linearity of the log of uptake vs. the ionization potentials was interpreted as a strong evidence for coordination bonding of these collagen matrices. The bonding may be primarily to uncharged amino, imidazole, or guanidino groups in the substrates.

Others have suggested that the metal coordination complex formation in cartilage,[26] elastin,[27] and egg-shell matrix[22] is important in the primary calcification process. Similar complexes may be involved in the calcification in bone and tendon matrices.

Wool is a complex protein which contains hydroxyl, amino, amide, carboxyl, sulphydryl, and disulphide groups. The amino groups are derived from arginine, lysine, and hystidine and the carboxyl groups from glutamic and aspartic acids. By titration and analysis, wool has been shown to have acid and base binding capacities both of about 0·8 meq/g: these capacities are in the range of those of synthetic ion-exchange resins,

TABLE 6.4. CONCENTRATIONS OF IONS BOUND TO COLLAGEN SAMPLES

Experimental conditions during exposure to ionic solutions

	Experiment 1	Experiment 2
pH	1 + 6·3, 2 + 6·4, 3 + 3·5 } Adjusted NH_4OH	5·4; Adjusted NaOH
Ionic strength	0·011–0·016 M	0·16 M
Indiv. ion conc.	0·3–3·0 mM	4·0 mM
Sample/solution (w/v)	0·05 g/100 ml.	0·2 g/100 ml.
Exposure	3 days	2 days

	Demin. bone	Demin. tendon	Washed tendon	Demin. bone	Demin. tendon
Li	—			?	?
Na	—			—	+
Ag	++++			+++	++++
K	—			—	+
Rb	—			—	+
Cs	—			—	+
Be	+	+		+++	++
Mg	+	+		+	+
Cu	++	++		++++	+++
Zn	++	+		++++	+
Mn	+	+			
Cd	+	—		+	+
Hg	—	—			
Sr	—	—		+	++
Pb	++++	+		++++	++++
Ba	—	—		+	+
Ca				+	+
Co				+	+
Ni				+	+
Al	+		++		
Ga	+		+++		
Fe	+		++++		
Cr	+		+++		
Bi	+		++++		

For key to symbols in Table 6.4 *See p. 238*

TABLE 6.4. *(cont.)*

—	less than 0·1 μmoles/g
+	0·1–10 μmoles/g
+ +	10–30 μmoles/g
+ + +	30–50 μmoles/g
+ + + +	50 or more μmoles/g

(From J. A. Spadaro *et al.*, *Nature, Lond.* **225**, 1134 (1970).)

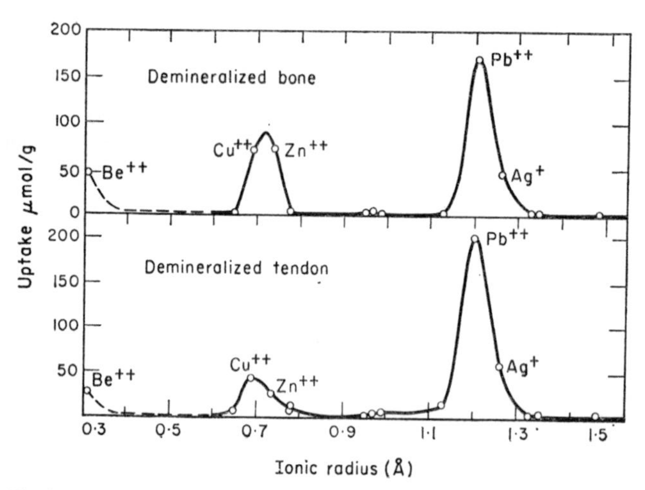

FIG. 6.3. The ion uptake (μmoles/g dry weight of tissue) as a function of ionic radius for demineralized bone and demineralized tendon in experiment 2. Trivalent ions were omitted in experiment 2 to maintain the pH of the hetero-ionic solutions above 5. The two preferred size regions are evident. (From J. A. Spadaro *et al., Nature, Lond.* **225**, 1134 (1970).)

and two papers dealing with chromatography on protein thin layers obtained from wool have been published.[28, 29]

Wool has in fact been shown to have ion-exchange capability[30] but because of its fibrous form, it has not been convenient and effective for use in chromatographic procedures. A more convenient form is prepared by rendering fibres into component cortical cells, and thin layers and columns have been prepared from them.[31] Keratin layers could be removed from the glass surface and recovered intact after brief immersion

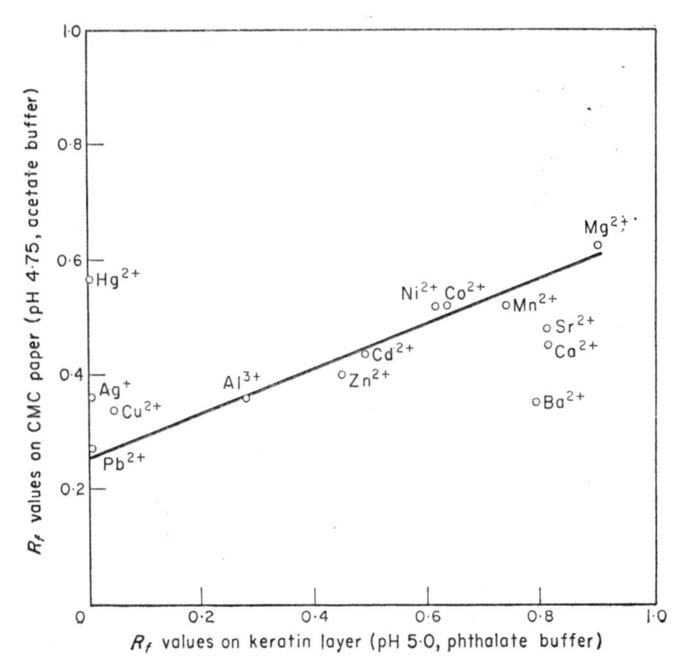

FIG. 6.4. Comparison of R_f values of metal ions on carboxymethylcellulose paper in acetate buffer at pH 4·75 and on keratin in phthalate buffer at pH 5·0. (From P. R. Brady *et al.*, *J. Chromatogr.* **54,** 55 (1971).)

of the plate in hot water (2 min at 70 °C). After drying, the keratin film had a papery texture and retained its characteristic strength and flexibility.

The migration of many of the metal ions was similar to that observed by Lederer[32] on carboxymethylcellulose paper, as illustrated in Fig. 6.4 in which the R_f values of the cations in buffer at pH 5·0 on unmodified keratin layers listed in Table 6.5 are plotted against those at pH 4·75 on carboxymethylcellulose paper. A linear correlation between the R_f values of the two types of substrate is evident for all but Cu, Hg, Ag, and alkaline earth metals. Both Hg and Ag were strongly retained by keratin probably because they reacted with cystine disulphide bonds present at 420 μmol cystine gram^{-1}. Cu exhibited a remarkable tailing. Ca, Sr, and Ba were not satisfactorily resolved. Na and K had little or no measurable affinity to wool and keratin.

TABLE 6.5. R_f VALUES OF CATIONS ON THIN KERATIN LAYERS

Cation	R_f values: Unmodified keratin B	Unmodified keratin A	Unmodified keratin C	Deaminated keratin B
Ag^+	×*	×	0·03T	—
Pb^{2+}	×	×	0·04T	—
Hg^{2+}	×	×	0·00	—
Fe^{3+}	×	×	0·00	—
Cu^{2+}	0·09T	0·04T	0·05T	—
Cd^{2+}	0·59	0·49	0·22	0·16
Co^{2+}	0·72	0·63	0·47	0·22
Ni^{2+}	0·78	0·62	0·43	0·28
Mn^{2+}	0·84	0·74	0·49	0·32
Zn^{2+}	0·64	0·40	0·28	0·17
Cr^{3+}	0·47	0·30	0·20	—
Mg^{2+}	0·91	0·90	0·80	0·56
Al^{3+}	0·52	0·28	0·15	0·06T
Ca^{2+}	0·90	0·82	0·65	0·44
Sr^{2+}	0·89	0·82	0·63	0·44
Ba^{2+}	0·84	0·80	0·64	0·39
Sn^{4+}	0·05	0·04	0·02	—

The solvent buffers were : (A) phthalate buffer, pH 4·0, ionic strength 0·2; (B) phthalate buffer, pH 5·0, ionic strength 0·2; (C) 0·2 M KNO_3 adjusted to pH 5·0 with nitric acid.

* × indicates that these cations precipitate in phthalate buffer.

T indicates tailing of spots.

(From P. R. Brady *et al.*, *J. Chromatogr.* **54**, 55 (1971).)

Pyrimidines and purines as well as dinitrophenylamino acids were chromatographed on keratin layers.[29] Some results are reported in Table 6.6.

Nucleic acids are known to interact with various divalent metal ions.[33–42] In modern biochemical and biophysical research, attention has been paid to the specific changes of the DNA secondary structure induced by copper ion binding.[43–52] The fact that copper decreases the melting

TABLE 6.6. R_f VALUES OF PYRIMIDINES AND PURINES ON METHYL-ESTERIFIED PROTEIN LAYERS

Compound	R_f values 100×	
	S_4	S_5
Pyrimidine		
4-Amino-5-bromo-	83	91
4-Amino-5-bromo-2-hydroxy-	57	74
4-Amino-2,6-dihydroxy-5-nitro-	3	34
4-Amino-2-mercapto-	55	67
4,5-Diamino-	66	77
4,5-Diamino-2,6-dimercapto-	10	45
4,5-Diamino-6-hydroxy-	57	70
4,5-Diamino-6-hydroxy-2-mercapto-	28	61
4,5-Diamino-6-mercapto-	51	67
4,5-Diamino-6-methylmercapto-	77	85
4,6-Dihydroxy-5-nitro-	4	10
2-Hydroxy-4-mercapto-	63	72
Purine		
2-Amino-	74	
6-Amino-	59	
2-Hydroxy-	34	
6-Mercapto-	60	
8-Mercapto-	68	
Purine	81	
8-Azaadenine	52	

R_f values are means of three determinations. Solvent systems are: S_4 = *n*-butanol saturated at 25° with water; S_5 = *n*-butanol–water–glacial acetic acid (70 : 20 : 10). (From P. R. Brady *et al.*, *J. Chromatogr.* **54**, 65 (1971).)

temperature of the DNA molecule has been attributed to the formation of a complex between copper and the nitrogen bases during heating. Besides the phosphate sites, guanine as well as cytosine[53] have been regarded as preferred binding sites in the formation of DNA–Cu complex. The denaturation of DNA at low ionic strength in the presence of copper ions has been shown to be reversible with increasing ionic strength.[43, 44] These results led to the assumption that the copper ion is complexed between guanine and cytosine.[43]

Eichhorn and Clark have postulated a model for the copper-denaturated state of DNA in which the unwound strands are held in alignment by formation of a copper complex between guanine and cytosine.[43] By using the electrolyte-induced reversion reaction, it was possible to show that copper cross links between the opposite strands are sensitive to

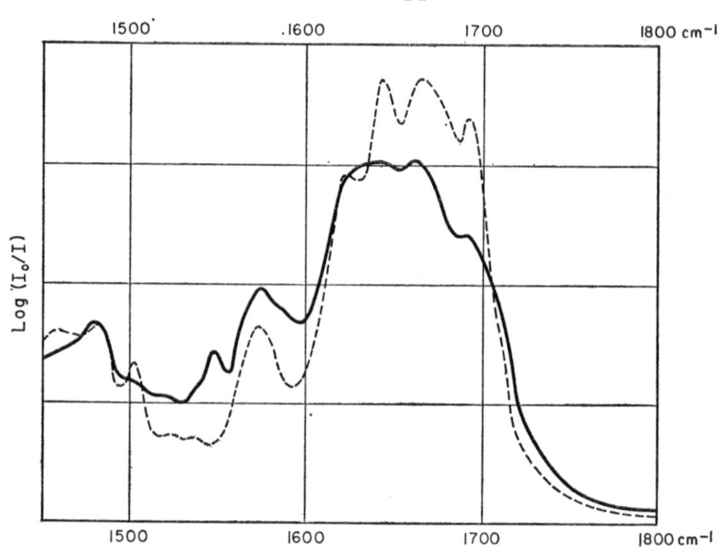

FIG. 6.5. Infrared spectra of DNA-Cu^{+2} complexes from *S. maxima* DNA (20 mole-%, film of DNA in D_2O atmosphere at 86% RH): (–) without Cu^{2+}; (——) 1 Cu^{2+}/DNA. (From Zimmer *et al.*, *Biopolymers* **10**, 441 (1971).)

temperature and to copper concentration.[54] The reversion reaction depends on the GC (guanine and cytosine) content. AT-rich DNA shows practically no reversion. Therefore, one is tempted to conclude that a considerable number of copper bridges are necessary for the re-formation process. This high affinity of copper to GC bases is in agreement with the increased tendency of GC-rich DNA to react with metal ions.

The infrared spectra of DNA-Cu complexes in Fig. 6.5, exhibit binding of copper to guanine and cytosine in DNA, similar to the spectra of the monomers.[53] This also shows the high affinity of the bases guanine and cytosine to copper in addition to its interaction with phosphate. The latter is concluded from an increase in the melting temperature at extremely low copper-base ratios and an increase in the sedimentation coefficient.[43]

Regarding the binding sites of the monomers as derived from n.m.r.[49] and i.r.[54] measurements, it was concluded that the copper ion cross-links involve N-7 of guanine and N-3 of cytosine as the most probable coordination sites (Fig. 6.6). The complexing ability of amino groups in

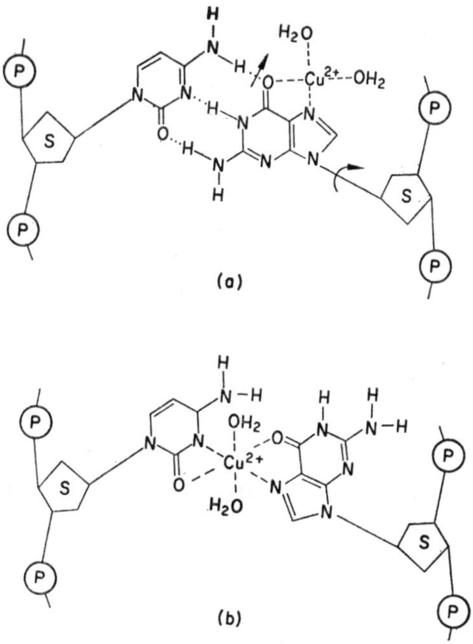

FIG. 6.6. Schematic representation of a complex binding model of Cu^{2+} in DNA at a GC base base pair: (a) most probable first attachment sites; (b) complexing of Cu^{2+} between guanine and cytosine moieties. (From Ch. Zimmer *et al.*, *Biopolymers* **10,** 441 (1971).)

pyrimidines with copper is low. In corresponding synthetic polynucleotides containing aminomethylated cytosine,[55] thimine[56] or uridine,[57] no complexation with copper could be detected. The main binding position N-3 of cytosine was recently deduced from studies of the crystal structure of the cytosine–copper complex.

FATTY ACID DERIVATIVES

Cutins are materials which form the cuticular membranes of the plants. The structures of the cutins are still incompletely known. They appear to be cross-linked polymers formed from bi- and trifunctional hydroxy fatty acids. Their characteristic properties, which permit them to provide plants with glassy, insoluble external coatings, are typical of cross-linked polymers. Cutins from a number of plants have been studied chemically: alkaline hydrolysis yields mixtures of hydroxylated fatty acids, some of which have been identified.[58]

Of course, the information available at present on the interactions of metal ions with cutin are very scarce; however, on the basis of the collection percentages listed in Table 6.7 it can be said that cutin shows a

TABLE 6.7. TRANSITION AND POST TRANSITIONAL METAL ION COLLECTION ON 200 mg CUTIN. PER CENT OF THE AMOUNT OF METAL PRESENT IN 50 ml OF 0·44 mM AQUEOUS SOLUTION (ATOMIC ABSORPTION SPECTROMETRY)

pH	hr	Cr(III)	Cr(VI)	Mn(II)	Fe(II)	Ni(II)	Cu(II)	Zn(II)	As(V)
2·5	1	36	28	20	16	30	81	0	80
	12	22	30	19	12	30	42	7	70
5·5	1		24	48	16	60	24	43	80
	12		16	51	25	64	69	46	80

pH	hr	Mo(VI)	Ag(I)	Sn(II)	Sb tartrate	Hg(II)	Pb(II)
2·5	1	73	63	100	44	50	32
	12		27	100	40	42	27
5·5	1	0	85	100		60	80
	12		50			60	77

(R. A. A. Muzzarelli, original results.)

behaviour which recalls that of CM cellulose (compare Table 6.7 with Tables 2.20 and 2.21).

Cutins would probably deserve more research on their characterization as chelating polymers.

REFERENCES

1. A. HAUG and O. SMIDRØD, *Acta Chem. Scan.* **24,** 843 (1970).
2. H. H. G. JELLINEK and S. P. SANGAL, *Water Res.* **6,** 305 (1972).
3. R. KOHN and I. FURDA, *Coll. Czech. Chem. Comm.* **32,** 1925 (1967).
4. R. KOHN, *Zucker* **21,** 420 (1968).
5. R. KOHN, *Zucker* **21,** 468 (1968).
6. R. KOHN and I. FURDA, *Coll. Czech. Chem. Comm.* **23,** 2217 (1968).
7. R. KOHN and V. TIBENSKY, *Coll. Czech. Chem. Comm.* **36,** 92 (1971).
8. A. BRZESKI, R. PAUL and W. SCHMEDDING, Brit. patent 1,097,380; 3 Jan. 1968.
9. L. S. HERS, H. H. WEETALL and I. W. BROWN, *J. Biomed. Mater. Res. Symp.* **1,** 99 (1970).
10. H. KIMIZUKA, A. YAMAUCHI and T. MORI, *Bull. Chem. Soc. Japan* **40,** 1281 (1967).
11. A. YAMAUCHI and H. KIMIZUKA, *Mem. Fac. Sci. Kyushu Univ.* Ser. C, **6,** 1 (1967).
12. M. M. KONONOVA, *Soil Organic Matter*, Pergamon Press, Oxford (1966).
13. J. P. MARTIN, *Soil Biol. Biochem.* **3,** 33 (1971).
14. D. S. LEHMAN, *Soil Sci. Soc. Proc.* 167 (1963).
15. J. L. MORTENSEN, *Soil Sci. Soc. Proc.* 179 (1963).
16. J. P. MARTIN and S. J. RICHARDS, *J. Bacteriol.* **85,** 1288 (1963).
17. J. P. MARTIN and S. J. RICHARDS, *Soil. Sci. Soc. Amer. Proc.* **33,** 421 (1969).
18. S. A. VISSER and H. MENDEL, *Soil Biol. Bioch.* **3,** 259 (1971).
19. M. DJURICIC, R. C. MURPHY, D. VITOROVIC and K. BIEMANN, *Geoch. Cosmoch. Acta* **35,** 1201 (1971).
20. F. J. STEVENSON and K. M. GOH, *Geoch. Cosmoch. Acta* **35,** 471 (1971).
21. M. T. DOIG and D. F. MARTIN, *Water Res.* **5,** 689 (1971).
22. A. WEINSTOCK, P. C. KING and R. E. WUTHIER, *Biochem. J.* **102,** 983 (1967).
23. K. SIMKISS and C. TYLER, *Quart. J. Microsc. Sci.* **99,** 5 (1958).
24. J. BELLO, *Bioch. Biophys. Acta* **109,** 250 (1965).
25. J. A. SPADARO, R. O. BECKER and C. H. BACHMAN, *Nature, Lond.* **225,** 1134 (1970).
26. J. R. DUNSTONE, *Biochem. J.* **72,** 465 (1959).
27. E. SCHIFFMAN, B. A. CORCORAN and G. R. MARTIN, *Arch. Biochem. Biophys.* **115,** 87 (1966).
28. P. R. BRADY and R. M. HOSKINSON, *J. Chromatogr.* **54,** 55 (1971).
29. P. R. BRADY and R. M. HOSKINSON, *J. Chromatogr.* **54,** 65 (1971).
30. R. KUNIN, *Ion Exchange Resins*, Wiley, New York, p. 74 (1958).
31. P. R. BRADY, J. DELMENICO and R. M. HOSKINSON, *J. Chromatogr.* **38,** 540 (1968).
32. M. LEDERER, *J. Chromatogr.* **29,** 306 (1967).

33. G. Zubay and P. Doty, *Biochim. Biophys. Acta* **29**, 47 (1958).
34. J. Shack and B. S. Bynum, *Nature, Lond.* **184**, 635 (1959).
35. E. Frieden and J. Alles, *J. Biol. Chem.* **230**, 797 (1958).
36. G. Felsenfeld and S. L. Huang, *Biochim. Biophys. Acta* **51**, 19 (1961).
37. T. Yaman and N. Davidson. *J. Am. Chem. Soc.* **83**, 2599, (1961).
38. G. Felsenfeld, *Symposium on Molecular Basis of Neoplasia*, p. 104. Univ. Texas Press, Austin, 1962.
39. S. Katz, *Nature, Lond.* **194**, 569 (1962).
40. S. Katz, *Nature, Lond.* **195**, 997 (1962).
41. G. L. Eichhorn, *Nature, Lond.* **194**, 474 (1962).
42. H. Venner and C. Zimmer, *Naturwiss.* **51**, 173 (1964).
43. G. L. Eichhorn and P. Clark, *Proc. Nat. Acad. Sci. U.S.* **47**, 778 (1965).
44. S. Hiai, *J. mol. Biol.* **11**, 672 (1965).
45. H. Venner and C. Zimmer, *Monatsber. Deut. Akad. Wiss.* **7**, 318 (1965).
46. J. H. Coates, D. O. Jordan and V. K. Srivastava, *Biochem. Biophys. Res. Commun.* **20**, 611 (1965).
47. V. I. Ivanov and L. E. Minchenkova, *Biochimia* **30**, 1233 (1965).
48. H. Venner and C. Zimmer, *Biopolymers* **4**, 321 (1966).
49. G. L. Eichhorn, P. Clark and E. D. Becker, *Biochemistry* **5**, 245 (1966).
50. K. B. Yatsimirskii, E. D. Kriss and T. I. Akhrameeva, *Dokl. Akad. Nauk. SSSR* **168**, 840 (1966).
51. S. E. Bryan and E. Frieden, *Biochemistry* **6**, 2728 (1967).
52. E. Safert and H. Venner, *Z. Physiol. Chem.* **340**, 157 (1965).
53. W. W. H. Fritzsche and C. Zimmer, *Europ. J. Biochem.* **5**, 42 (1968).
54. C. Zimmer, G. Luck, H. Fritzsche and H. Triebel, *Biopolymers* **10**, 441 (1971).
55. C. Zimmer and W. Szer, *Acta Biochim. Polon.* **15**, 339 (1968).
56. W. Szer, *Acta Biochim. Polon.* **13**, 251 (1966).
57. G. L. Eichhorn and E. Tarien, *Biopolymers* **5**, 273 (1967).
58. T. A. Geissman, *Principles of Organic Chemistry*, p. 822, Freeman & Co. Publ., London, 1968.

INDEX

OTHER TITLES IN THE SERIES IN ANALYTICAL CHEMISTRY

Vol. 1. WEISZ—Microanalysis by the Ring Oven Technique.
Vol. 2. CROUTHAMEL—Applied Gamma-ray Spectrometry.
Vol. 3. VICKERY—The Analytical Chemistry of the Rare Earths.
Vol. 4. HEADRIDGE—Photometric Titrations.
Vol. 5. BUSEV—The Analytical Chemistry of Indium.
Vol. 6. ELWELL and GIDLEY—Atomic Absorption Spectrophotometry.
Vol. 7. ERDEY—Gravimetric Analysis Parts I-III.
Vol. 8. CRITCHFIELD—Organic Functional Group Analysis.
Vol. 9. MOSES—Analytical Chemistry of the Actinide Elements.
Vol. 10. RYABCHIKOV and GOL'BRAIKH—The Analytical Chemistry of Thorium.
Vol. 11. CALI—Trace Analysis for Semiconductor Materials.
Vol. 12. ZUMAN—Organic Polarographic Analysis.
Vol. 13. RECHNITZ—Controlled-potential Analysis.
Vol. 14. MILNER—Analysis of Petroleum for Trace Elements.
Vol. 15. ALIMARIN and PETRIKOVA—Inorganic Ultramicroanalysis.
Vol. 16. MOSHIER—Analytical Chemistry of Niobium and Tantalum.
Vol. 17. JEFFERY and KIPPING—Gas Analysis by Gas Chromatography.
Vol. 18. NIELSEN—Kinetics of Precipitation.
Vol. 19. CALEY—Analysis of Ancient Metals.
Vol. 20. MOSES—Nuclear Techniques in Analytical Chemistry.
Vol. 21. PUNGOR—Oscillometry and Conductometry.
Vol. 22. J. ZYKA—Newer Redox Titrants.
Vol. 23. MOSHIER and SIEVERS—Gas Chromatography of Metal Chelates.
Vol. 24. BEAMISH—The Analytical Chemistry of the Noble Metals.
Vol. 25. YATSIMIRSKII—Kinetic Methods of Analysis.
Vol. 26. SZABADVÁRY—History of Analytical Chemistry.
Vol. 27. YOUNG—The Analytical Chemistry of Cobalt.
Vol. 28. LEWIS, OTT and SINE—The Analysis of Nickel.
Vol. 29. BRAUN and TÖLGYESSY—Radiometric Titrations.
Vol. 30. RUŽIČKA and STARÝ—Substoichiometry in Radiochemical Analysis.
Vol. 31. CROMPTON—The Analysis of Organoaluminium and Organozinc Compounds.
Vol. 32. SCHILT—Analytical Applications of 1,10 Phenanthroline and Related Compounds.
Vol. 33. BARK and BARK—Thermometric Titrimetry.
Vol. 34. GUILBAULT—Enzymatic Methods of Analysis.
Vol. 35. WAINERDI—Analytical Chemistry in Space.
Vol. 36. JEFFERY—Chemical Methods of Rock Analysis.
Vol. 37. WEISZ—Microanalysis by the Ring-Oven Technique. (2nd Edition—enlarged and revised.)

Vol. 38. RIEMAN and WALTON—Ion Exchange in Analytical Chemistry.
Vol. 39. GORSUCH—The Destruction of Organic Matter.
Vol. 40. MUKHERJI—Analytical Chemistry of Zirconium and Hafnium.
Vol. 41. ADAMS and DAMS—Applied Gamma Ray Spectrometry (Second edition).
Vol. 42. BECKEY—Field Ionization Mass Spectrometry.
Vol. 43. LEWIS and OTT—Analytical Chemistry of Nickel.
Vol. 44. SILVERMAN—Determination of Impurities in Nuclear Grade Sodium Metal.
Vol. 45. KUHNERT-BRANDSTATTER—Thermomicroscopy in the Analysis of Pharmaceuticals.
Vol. 46. CROMPTON—Chemical Analysis of Additives in Plastics.
Vol. 47. ELWELL and WOOD—Analytical Chemistry of Molybdenum and Tungsten.
Vol. 48. BEAMISH and VAN LOON—Recent Advances in the Analytical Chemistry of the Noble Metals.
Vol. 49. TÖLGYESSY, BRAUN and KYRS—Isotope Dilution Analysis.
Vol. 50. MAJUMDAR—N-Benzoylphenylhydroxylamine and its Analogues.
Vol. 51. BISHOP—Indicators.
Vol. 52. PRIBIL—Analytical Applications of EDTA and Related Compounds.
Vol. 53. BAKER and BETTERIDGE—Photoelectron Spectroscopy: Chemical and Analytical Aspects.
Vol. 54. BURGER—Organic Reagents in Metal Analysis.